Henry Herbert Donaldson

The Growth of the Brain

A Study of the Nervous System

Henry Herbert Donaldson

The Growth of the Brain
A Study of the Nervous System

ISBN/EAN: 9783744723800

Printed in Europe, USA, Canada, Australia, Japan

Cover: Foto ©berggeist007 / pixelio.de

More available books at **www.hansebooks.com**

THE GROWTH OF THE BRAIN:

A STUDY OF THE NERVOUS SYSTEM IN RELATION TO EDUCATION.

BY

HENRY HERBERT DONALDSON,

Professor of Neurology in the University of Chicago.

THE WALTER SCOTT PUBLISHING CO., LTD.
PATERNOSTER SQUARE, LONDON, E.C.
CHARLES SCRIBNER'S SONS
597 FIFTH AVENUE, NEW YORK
1914

PREFACE.

WE are told that this age is one of nervous strain. Probably in their own time the monsters of cretaceous days would have expressed a similar opinion, had the opportunity been granted them. From the beginning, the outer world has modified all animals possessed of a nervous system mainly by its aid, and as far as this system could alter its reactions, just so far could the animal adapt itself to the changed environment.

In this generation, to be sure, our cephalic centres are sometimes overworked, whereas in the remote past the stress fell more on other parts; but we are rather allied to the rock-bound dead by an inherited power to respond than separated from them by a recent capacity for nerve exhaustion.

As the reader will perceive, these remarks might serve to magnify the office of this book by suggesting how fundamental to the welfare of all higher animals are the powers of the nervous system, and therefore how important it will be to search out the growth changes which produce them.

It seemed desirable to bring together in a comprehensive way the facts bearing on this portion of the problem. In discussions upon growth, the valuable records on brain-weight are sometimes alone brought forward, but there exist a vast number of other facts, which, when joined with these, illuminate not only them, but the entire field of view, and indicate the unworked areas within it. I have therefore sought especially to emphasise some more neglected points. Let me enumerate a few: the growth of the nervous system compared with that of the body; the interpretation of brain-weight in terms of cell structure; the early limitation of the number of nerve cells; the peculiar relation in this system between increase in size and in organisation; the large though variable number of cells which have but slight importance in the final structure; the dominance of nutritive conditions; the wide diffusion of nerve impulses; the incompleteness of repose; the reflex nature of all responses; the native character of mental powers; and the comparative insignificance of formal education.

To those who deal with the nervous system as it grows, the facts within these covers will, I trust, be useful; the parent, the teacher, and the physician represent three classes who seek for light upon such matters, and it would be fortunate if, as a result of their demands, there soon should be supplied an account of growth far more extensive and more luminous than this.

In putting together the materials employed, I have sought the counsel and assistance of my friends who took an interest in the work, and to them I would express my gratitude, yet without transferring my responsibilities. Save in a few cases, the illustrations in this book have been copied from standard works, and I am greatly indebted to both authors and publishers for the uniform courtesy with which permission has been granted to have this done. To Messrs. Macmillan & Co., publishers of Foster's *Physiology;* Longmans, Green & Co., publishers of Quain's *Anatomy;* Ginn & Co., publishers of *The Journal of Morphology;* D. Appleton & Co., publishers of *The Popular Science Monthly*, and to the proprietors of *The Lancet*, my acknowledgments are especially due.

CHICAGO, *May*, 1895. H. H. D.

CONTENTS.

PAGE

CHAPTER XIV.

CHAPTER XV.

CHAPTER XVI.

CHAPTER XVII.

LIST OF ILLUSTRATIONS.

LIST OF TABLES.

CHAPTER I.

AN INTRODUCTION TO THE STUDY OF GROWTH.

Problems—Cells—Egg-cell—Fertilisation—Cell-multiplication and cell-enlargement—Number of cells in man—Nutrition of cells—Specialisation—Germ layers—Methods of growth—Cells not simple—Theory of development—Relations between nucleus and cytoplasm—Entire larvæ from fractions of the egg—Multiple embryos from a single egg—Size of animals—Growing cells—Specialisation antagonistic to regeneration—Longevity—Rate of growth—Onset of decay—Two habits of growth—Cell-multiplication occurs early—Time limitations of development—Size in geological time—Conditions influencing size—Rhythm of growth—Summary—Outline of succeeding chapters.

THE living world of which we are a part is ever changing. Unceasingly from seeds and eggs new generations arise, and each day those which have completed the cycle of a life pass to their final dissolution. The history of these changes forms a record difficult to interpret, yet perennially interesting, since each question solved is at once replaced by the group of problems from which it sprang. Among the current problems I shall select a series, and the study here proposed will deal with those relating to the growth of the brains of animals. To the understanding of them, a review of the more important observations which broadly underlie the laws of growth will form a fitting introduction.

The higher animals possessed of a backbone constitute the vertebrate group, to which man himself belongs, and since the changes in the vertebrates are those of greatest interest, the facts presented will especially apply to them.

Minute anatomy has taught us that all animals are composed of small structural elements—cells. Those microscopic portions of living matter, usually but a small fraction of a millimeter in diameter, possess, whatever their shape, two principal parts: a small and denser portion centrally located, the nucleus, and surrounding this a more fluid and bulkier portion, which forms the body of the cell, the cytoplasm; the latter being so altered at its surface that its outermost layer constitutes an enclosing envelope, the cell-membrane. The cytoplasm and the enclosed nucleus are only the most evident and not the sole constituents in the typical cell, but the finer anatomy of these elements need not be here described.

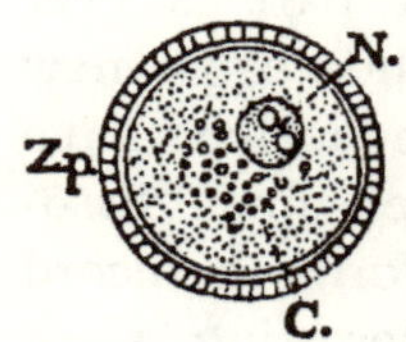

FIG. 1.—Unfertilised human ovum, × 170 diam. (Nagel). C. Cytoplasm; N. Nucleus; Z.p. Zona pellucida.

These are the units by the production of which the growing animal is formed, and of which the mature animal is principally composed, and therefore all of them are the lineal descendants of the ovum, which is, in its first form, but a single cell. The formation of the animal-body from this simplest beginning, with all that this implies of cell-multiplication, enlargement, and specialisation, is the result of changes first showing themselves within the ovum. (Fig. 1.)

In order that these changes may occur, the ovum must, save in a few exceptional cases, undergo fertilisation, the essential feature of which is the union with it of a second element, the sperm-cell. As a result of this union, the cytoplasm of the two cells is condensed into one, the two nuclei unite to form a single nucleus, and thus modified, the fertilised ovum or egg begins to grow.

The first indication of such growth is a division of the egg into a number of cells, without much enlargement of the total mass.

This is apparently the arrangement of the cell substance best suited for vigorous nutrition, and as the increase in bulk depends on nutritive processes, the importance of it is plain.[1] Later the nutritive conditions

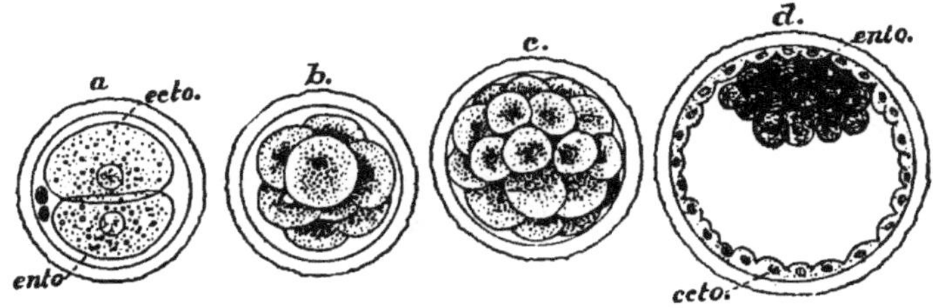

FIG. 2.—Segmentation of a mammalian ovum. *a*, *b*, *c*, semi-diagrammatic × 170 diam. (Allen Thomsen, after v. Beneden, Quain's *Anatomy*) ; *d*. Section of the ovum of the rabbit during the later stages of segmentation (E. v. Beneden, Quain's *Anatomy*) ; *ecto*=ectoderm ; *ento*=entoderm.

improve, and with this improvement the size of the mass increases.

The processes which lead to subdivision are first evident in the nuclear structures, which become the seat of complicated but, at the same time, perfectly orderly changes. As a result, the constituents of the nucleus divide into two portions which separate from one another, the surrounding cytoplasm gathers about the respective daughter-nuclei, and these two portions then become marked off by the cell-membrane; thus, in the place of one cell, two cells are now present. By cell-

[1] Sachs, *Flora*, 1893.

division occurring in this way, and accompanied by changes in the size and form of the elements, the complex individual is built up.

What is thus accomplished in the case of man is roughly indicated by the following calculations:—According to the estimate of C. Francke,[1] there are, in the entire body of a full-grown person, a number of fixed cells represented by 4,000,000,000,000. The cells in the blood are not fixed, and for these the best calculations give 22,500,000,000,000, or, in the entire body, a total of 26,500,000,000,000 cells. This number may appear surprisingly large, but it must be remembered that there are in a cubic millimeter 1,000,000,000 cubic μ (the micron $= \mu$ being 0·001 of a millimeter, and the unit adopted by histologists in micrometry); so that if each cell in the human body had but the volume of one cubic μ, the whole number could be condensed into 22,500 cu. mm., or 22·5 cu. cm., not a very large mass, since it is represented by a cube length whose edge is 2·8 cm. (Fig. 3.)

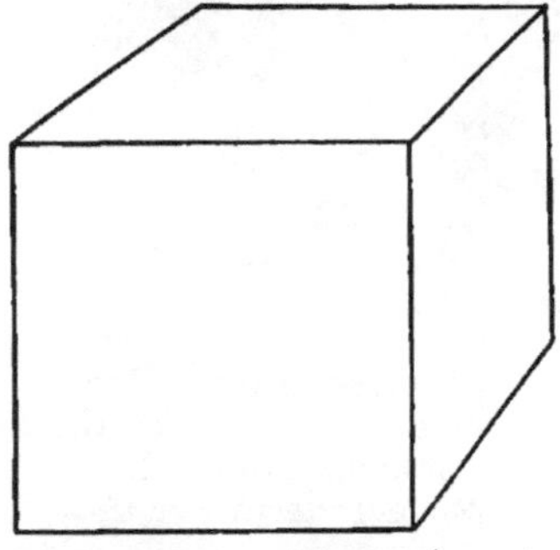

FIG. 3.—Representing a cube 2·8 cm. long on each edge. This cube could contain 26,500,000,000,000 cells, provided each cell had a volume of 1 cubic micron, the micron being the unit of measure employed by histologists.

These facts can be expressed in still another form which will indicate the average size of the cells. A man weighing 150 lbs.—the average specific gravity of his body being taken as 1·030—would have a volume of 66,200 cu. cms. As stated above, all the cells of the body might be contained in 22·5 cu. cm., provided each cell had a volume of only 1 cu. μ. By

[1] Francke, *Die menschliche Zelle*, 1891.

simple division of the former volume by the latter, it appears that the average volume of the cells under the condition named, is 2,942 cu. μ, equivalent to a cube 14 μ, on each edge, which is by no means small.

The nutrition of a cell depends on the surrounding medium from which it derives food, either directly by the incorporation of small particles, or, as is more usual in higher animals, by the diffusion of the nutrient substances in a fluid form. For this latter process the surface of the cell must be exposed to the nutrient fluid, a condition easily fulfilled in the case of single cells or those arranged in thin layers. When, however, the mass of cells becomes large, the nourishment of those below the surface is indirectly accomplished by the conveyance of the nutrient fluid to them through vessels that penetrate the mass. Between the blood in these vessels and the surrounding cells the process of diffusion then takes place. This arrangement leaves to the cells most directly in contact with the ingested material the peculiar duty of preparing the food for distribution by diffusion, and here arises a good instance of the division of labour among the cells leading to that mutual dependence of which the completed organism offers so many examples, for it holds true through the entire series of the activities of the animal body, that each is carried on by elements specially modified for the purpose.

All these different forms of physiological activity exhibited by the higher animals are found to be present in many of the unicellular organisms, and are assumed to be present in them all. For this reason, the development of different capacities by different groups of cells is best regarded, neither as the sudden acquisition of a new power, nor as a peculiar capacity only inherent in a particular group of elements, but as arising simply from an emphasis of one of the several powers originally common to them all. The specialised cell is therefore physiologically unbalanced, if not incomplete; and so the more exquisitely a cell is adapted to some particular function, the less capable it is of performing the entire series of reactions indispensable to its very existence; hence the more dependent it becomes upon its neighbours.

Very early in the history of the development of an animal, before any of the organs are formed, yet when the number of cells is already large, these latter become separated into layers. (Fig. 4.)

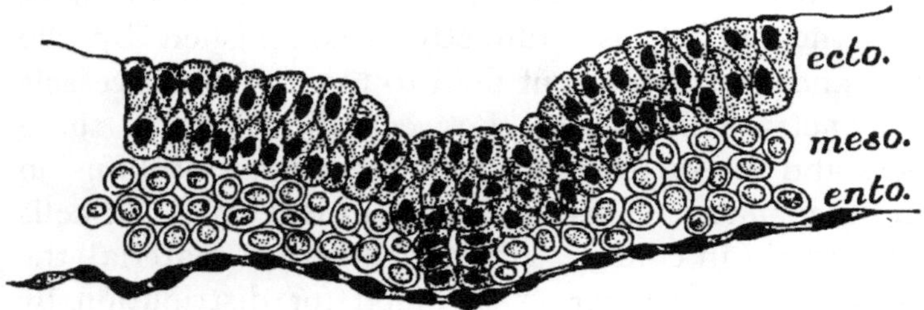

FIG. 4.—Transverse section to show the germ layers of a mole embryo, × 265 diam. (Heape, Quain's *Anatomy*); *ecto.*, ectoderm; *ento.*, entoderm; *meso.*, mesoderm.

The germ layers, as they are called, are composed of cell-groups, which tend to show special characters, according to the layer in which they occur. Ultimately the layers give rise to three main subdivisions of the body. From the ectoderm, or outer layer, are derived the skin and the nervous system; from the inner layer, the entoderm, the alimentary tract with its appendages; and from the layer which lies between the two, the mesoderm, are derived the muscles, the supporting tissues, the reproductive system, and the circulatory apparatus.

It is for us to inquire how these portions grow, and

by what means the tiny organs to which they first give rise become the bulky organs of the adult. This increase must be due either to the swelling of the cells that are already formed, or to the formation of new ones by divisions of the old ; and in every instance it is to be determined whether one or both of these processes have been at work, and what share in the result may be attributed to each of them.

The conception of a complicated animal body, as composed of a number of cells, was at first accompanied by the idea that the cells themselves were comparatively simple, and the simplicity of the cell was contrasted with the complexity of the whole organism. Unfortunately this notion of simplicity is misleading, for there are instances, the most striking being among the protozoa, in which a number of functions are carried on by different parts of the cytoplasm, although, according to definition, the entire animal forms but a single cell. But the facts in connection with which the idea of simplicity is least fit, are those relating to development. From the human ovum, for example, is gradually elaborated the complex adult, and it is difficult to escape the conviction that in some measure at least the complexity of the latter is represented in the structure of the former.

In carrying out this elaboration the course of events is mainly determined, not by the surrounding conditions, although these have some influence, but by conditions within the egg. The beginning of the growth process immediately follows fertilisation, and hence that may be looked upon as the initial impulse, but in what manner this impulse sets in motion the marvellous series of changes which are to result in the completed animal is not thereby explained. A few generations ago the development of animals was regarded as due to evolution ; evolution in the sense that within the substance of the egg there was present, in an extremely reduced form to be sure, the perfect and complete animal, to which the egg was destined to give rise ; and it was held that the changes accompanying development consisted in an enlargement of this miniature. In this gross form the theory was soon discarded, but it is still current in a modern guise. The animal is no longer considered to be pre-existent as a formed miniature, but it is assumed by many that condensed in this small bit of living matter are particles, molecules, if you please, which in some way have marked out for them a fixed course of development. In a sense, therefore, the future animal is still looked upon as pre-existent in the egg, since we know of no device by which any widely different species of animal can be made to grow from that particular cell. Yet despite the fact that the lines are so narrowly fixed, along which the eggs of different species develop, there are only slight cytological differences to be observed among those from different animals ; and we have by no means reached the point where, by the examination of an egg-cell alone, any detailed prediction can be made concerning the form into which it might develop. The theory just mentioned probably expresses many of the facts of development, but there are other facts, such as the production of several embryos from a single egg or complete embryos from fractions of the egg, which show it to be at least incomplete. The significance of these non-conforming observations will be best understood from a few examples.

It has been shown repeatedly that if from a cell the nucleus be removed, the cytoplasm ceases to grow, and shortly dies. When thus isolated, the nucleus also dies. Recently Wilson has studied the development of the eggs of the Lancelet, Amphioxus.[1] In this simple fish it is possible to separate from one another the first two cells formed by the division of the egg. The cells thus obtained continue to develop and go through a number of stages characteristic of the normal egg, but the larva is only half as large as the normal one, its component cells being reduced in size. This larva dies young, and whether it could be made to develop further is left for future experiment to decide. When the egg has already formed four cells, it is still possible in the same manner to separate one of these, which will also

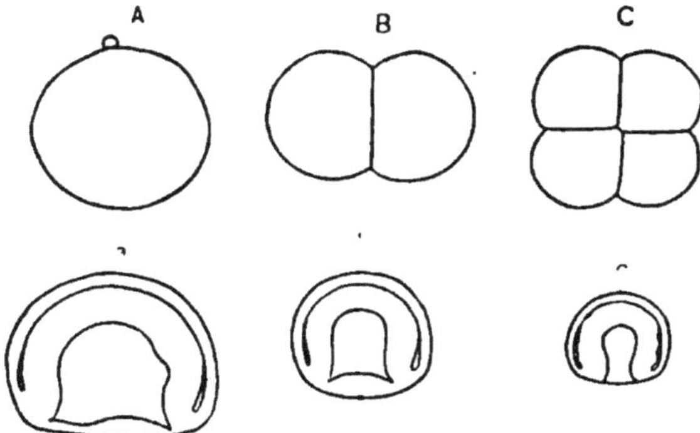

FIG. 5.—Showing the egg of Amphioxus in the one (A), two (B), and four cell (C) stages and below the corresponding larvæ, a, b, c developed in each case from a single cell. The larva becomes smaller as the fraction of the egg becomes less, × 135 diam. (Wilson.)

develop alone, and give rise to a complete larva, save that it is only one-quarter the normal size. This is still more difficult to rear (Fig. 5). During the past year it has been shown (Morgan) that fragments of the unsegmented egg of the sea urchin may, when they contain both the male and female pronuclei, develop gastulæ, and this occurs even if nucleated fragments have but one-fiftieth

[1] Wilson, *Journal of Morphology*, 1893.

the volume of the egg. The tiny gastula thus produced is composed of small cells, but not cells proportionately reduced in size, since they may have five times the volume of the cells composing the normal gastula.

Multiple embryos from a single egg have been obtained by Loeb[1] in yet a different way. This investigator has succeeded in causing eggs, just fertilised, but still undivided, to swell so that the enclosing membrane ruptured, and a portion of the cell contents extruded

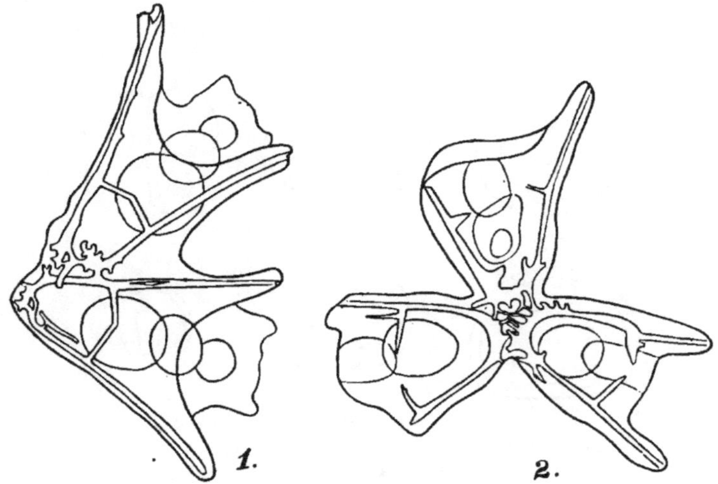

FIG. 6.—Multiple embryos from sea-urchin eggs, × 94 diam. (Loeb), 1 twin, 2 triplet.

in the form of one or more buds or droplets. In due season these buds, as well as the cell substance remaining within the membrane, developed embryos. The multiple embryos thus formed are represented in the accompanying figure. Thus, where at the start the egg was morphologically but a single cell, it has been found

[1] Loeb, vide *Biological Lectures, delivered at Wood's Holl.* Boston, 1894.

possible, through change in the conditions surrounding it, to cause the development within it of more than one embryo. (Fig. 6.)

In these experiments the attempt is made to grow an animal from a cell or a part of a cell, which represents only a fraction of the substance naturally intended for the entire organism, and as one consequence the embryo is below the normal size. To interpret such phenomena in harmony with the idea of preformation in the egg, is a present line of zoological endeavour; but, whatever may be the outcome, the striking fact remains, that such repair as occurs when complete embryos are reared from less than the entire egg, is not quantitative, because the size of the embryo bears a definite relation to the size of the original fragment.

In this connection it may be asked why animals grow to that particular size which is characteristic for the different species; why one egg yields a mouse, another an elephant? But until more facts on the size of egg-cells are available for comparison, speculation on the subject would be futile.

Young and growing cells have a nucleus which is large when compared with the amount of cytoplasm, and the possibilities of growth and division in animal cells appear to be dependent on this relation. If this condition is a necessary one, it should be found universally, even where growth, due to the regeneration of the lost parts, is taking place. For not only is the ovum capable of producing the one complete animal, but, in addition, retains, in many forms, the power to reproduce lost parts. This process is the more interesting in view of the fact that among the invertebrates generally, and in many lower vertebrates also, especially when young, the regeneration of the lost parts seems so simple and common an occurrence, and can take place on so magnificent a scale. By repeated removal of the limbs Spallanzani obtained from a single newt in the course of three months 687 new bones. In the adult animals, to be sure, the regenerated portion is at first small, and only gradually reaches the normal size, but it ultimately does so, and the ease of this process, the rapidity of it, as well as the number of times it can be repeated, are all highly suggestive. This capacity for the reproduction of parts lost is much diminished in the higher vertebrates, and the decrease in this power is explained by their higher specialisation, or, in other words, the greater the number of generations between any group of somatic cells and the ovum from which they are derived, the less the capacity in them for regeneration.

If the organism, as a whole, is highly specialised, it is due to the fact that the constituent cells are also specialised; so that in the animal series progressive specialisation may be expressed either in terms of the entire animal or its structural elements. It has just been indicated that the embryonic condition of cells is that most favourable to growth, and it therefore follows that the processes of regeneration would be least perfect in the higher vertebrates whose cells had departed furthest from the embryonic condition. It is recognised, moreover, that the growth changes in cells are but incompletely reversible. Thus, when a cell has once become in a high degree modified, it loses the capability of reverting, under new conditions, to its primitive state. Here, again, in the animal series a gradation is evident, for those cells which are least highly modified and those composing the less specialised animals are the ones in which this reversion is most readily accomplished. In a broad way, then, the capacity for regeneration implies a latent youthfulness.

That the power to grow is closely connected with the length of life in different animals is beyond question, but a proper formulation of this connection is difficult. Certainly there is as yet no adequate explanation of the fact that some animals live longer than others; we find that, in general, the larger animals are those having the longer life, and thus long life is to be associated with size; but, on examining this relation, it appears that size depends upon two factors—the rate of growth, and the length of time that growth continues, and animals that attain a large size grow both rapidly and for a long time. Minot has pointed out that if the guinea-pig,[1] rabbit, and man are compared as to their increase in weight the following interesting facts appear:—

TABLE I.—SHOWING THE AVERAGE DAILY INCREASE IN WEIGHT DURING THE GROWING PERIOD OF THE GUINEA-PIG, RABBIT, AND MAN. (*Minot.*)

Animal.	Weight at maturity in grms.		Number of days from birth to maturity.		Number of days of gestation.	Weight increase in grms. per diem.
Guinea-pigs	775	divided by	365	plus	67	1·82 per diem.
Rabbits ...	2,500	„	365	„	30	6·30 „
Man ...	63,000	„	9,139	„	289	6·69 „

This table shows that the rate of increase in the rabbit and in man is about the same; but owing to the fact that in man growth continues for about twenty-five years, whereas in the rabbit it stops at the end of one

[1] Minot, *Journal of Physiology*, 1891.

year, the difference in the final size of the animals is very great. Since, then, natural death is the result of old age, and old age is not attained until after growth is completed, and large size depends on long-continued growth, therefore long life and large size are found associated.

The rate of growth is measured by the proportional increase in a unit of time. This is well shown in the accompanying curve derived from Minot's observations on guinea-pigs.[1]

From the moment of birth, at least, there is a rapid

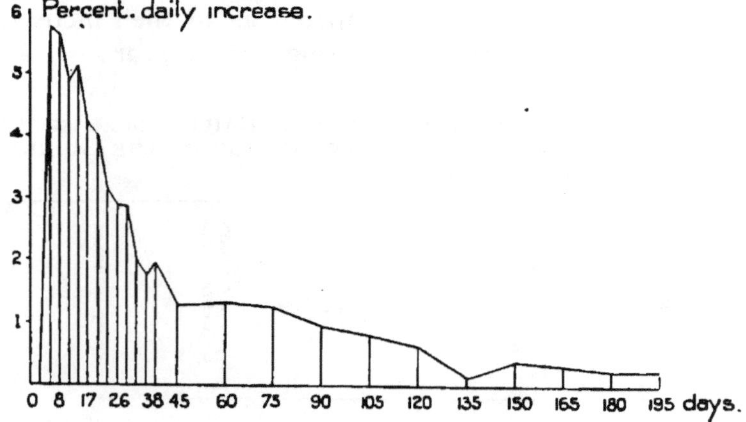

FIG. 7.—Curve to indicate in growing guinea-pigs the percentage of daily increase in weight, from birth to the 195th day of life. (Minot.)

falling off in the proportional daily increase in the body-weight. This is characteristic of all the higher animals, man included; so that from birth, and probably from the first division of the egg, the capacity to grow is undergoing an almost steady diminution. This capacity, as the curve shows, must soon reach a point where it

[1] Minot, *Proc. Am. Assoc.*, 1890.

practically amounts to nothing, and then it is that the body becomes most susceptible to destructive influences continuously at work upon it. With the decay of any portion of the organism comes a disturbance of the balance in the work allotted to the several portions. The other tissues not specialised for the purpose are forced to take on the functions of those already damaged, and this they can do but imperfectly. The result is progressive enfeeblement of these overtaxed cells, until finally some function is so imperfectly performed that the life processes are blocked, and the organism as a whole dies. Minot pointed out that in the mature cell the proportion of cytoplasm to nucleus was greater than in the growing animal; and on the ground of this propounds, half humorously, the aphorism, that "Protoplasm is the physical basis of advancing decrepitude." By thus granting that the cytoplasm is in some way a hindrance to its own activity, the specialised body-cells are seen to be similar to the unicellular organisms, among which cell-enlargement is so clearly self-limited. But it is plain that in speaking of cells and the relative development of their different portions, we are using these variations in the size of the nucleus and the cytoplasm merely as indices of more subtle structural and chemical changes which will form the proper basis for a future explanation.

Though we infer that the ability to grow is, so far as it goes, a protection against commencing deterioration, yet a search for illuminating details does not yield much because of our meagre information concerning the length of life in animals. Popularly birds, reptiles, batrachians, and fishes are considered capable of living for very many years. It further appears that some reptiles and some fishes continue to grow almost as long as they live, but in later life their growth is certainly very slow. At the same time birds, which in many instances are credited with more than a hundred years, have not been shown to undergo this long-continued growth. In the case of birds, therefore, neither continuous growth nor large size can be called in to explain long life, so that in reality the inherent conditions which determine longevity are still unrecognised.

Nevertheless vertebrates may be divided into two principal groups, according to the manner in which they grow. There are those in which the size of the individual is fixed, in the sense that some time long before the natural close of life growth ceases; and those in which growth changes are almost coextensive with life. To the former class man belongs, together with the other mammals and the birds, and to the latter some fish, amphibia, and reptiles.

It has been already stated that there are two ways in which an organism may increase in size: either by the increase of the number of cells composing it, or the number remaining the same, by an increase in their mass. As a matter of fact, the two sets of changes always go hand in hand, but the share they take in the final result is not the same in different animals or different tissues, nor is it the same at the various periods in the life history of the same animal. The topic is worthy of some elaboration.

The different classes of vertebrates are similar in the fact that during the early life of the embryo, before the organs are formed, the chief factor giving rise to the increase in size is the production of new cells. As this production becomes less rapid, those cells which have been already formed undergo enlargement, together with such changes in structure as fit them for their special duties. The first process is thus gradually replaced by the second. In the human nervous system the production of new cells ceases some time before birth, though in some other mammals it may go on for a longer period, but in all it very early becomes the less significant method of increase. Before attaining any functional importance these newly-formed cells must pass through a long series of developmental changes. Thus in any organism the total number of structural elements is greater, and in early life very much greater, than is the number functionally perfected. Two different methods in subsequent development of animals or tissues may be distinguished according to the time relations within which the constituent cells become mature: in the one case the newly-formed cells may change almost simultaneously, or in the other, these changes may be successive in the different cells, so that in the former instance the time interval demanded for the completion of growth is short, in the latter it is long. With these facts in view, an explanation of the growth processes in those animals which early attain a fixed size is to be found in the fact that in them cell development is condensed into a short period, which in the case of those in which growth is long continued the time during which the development of different cells may occur is increased. Development and the changes involved in growing old are, however, by no means synonymous, so that although in those animals with a fixed size there are always to be found undeveloped cells, yet it is not a correct inference that these cells are also young in the sense that they might still complete their development. It appears, rather, that the capacity for undergoing expansive change is transient, and that those cells which fail to react during the proper growing period of an animal have lost their opportunity for ever. Here again is a difference between the two groups of animals which have been contrasted, since in those the size of which is least limited by age the possibility of development is more persistent.

Connected with these facts are those bearing on the variations of size in animals of the same family or species. It is worthy of note that there are many cases where a type of vertebrate animal appearing at several geological horizons has the last representatives more bulky than the first, while the extremes are connected through the different horizons by a series of forms intermediate in size. Extinction in these cases apparently overtakes the type when its representatives have reached the greatest size, and the suggestion here is that species, like individuals, have a period of growth followed by specialisation, which latter is in the end the cause of their extinction, through the loss of adaptability to new conditions. On the other hand, some invertebrates slowly deteriorate in the course of their geological history—perhaps through an unequal struggle with changed conditions, which, though damaging, were not sufficient to exterminate them.

The influences affecting the animal from without are expressed by the term environment; while from within there is the complex of influences called heredity. Probably dependent on external conditions are the peculiarities of faunas on the smaller islands, one feature of which is the small size of the animals there found. So, too, the experiments of Yung[1] in rearing tadpoles indicate the relation between the capacity of the enclosing vessel and the possible size of the tadpole which was being reared in it. Semper and De Varigny obtained similar results with the pond snail (Lymnæa)[2], by growing them in different quantities of water, varying

[1] Yung, *Compt. Rend.*, 1885.

[2] Semper, *Animal Life*, 1881. De Varigny, *Journ. de l'anat. et physiol.*, Paris, 1894.

the conditions in many ways. Those in the larger amount of water were the larger, and beginning with those grown in the least quantity, which were the smallest, they could be arranged in a series increasing in size, according to the quantity of water in which they had been placed. The explanation of these suggestive results is, to say the least, obscure, but it must ultimately rest on variations in nutrition, which may be influenced by the amount of activity of the animal and the diffusion of its own waste products, which act as poisons, and which would be effective in proportion as the quantity of water was small. In a recent account of some conditions which modify growth, Bizzozero states that cell division is not stopped by starvation, though it may be thus decreased in intensity, and that inanition may even rouse dormant cells to the act of division. In the ear of the rabbit, hyperæmia causes increased cell multiplication, together with increase in size of the ear ; and the same result may be effected by high temperature. Conversely, low temperature retards in a most remarkable manner the increase of the growing ear. When, however, the low temperature is replaced by a normal degree of heat, the ear previously dwarfed overtakes the one which has been left under the usual conditions, and very soon becomes equal to it in size.[1] By the aid of the nervous system the response of an animal to conditions influencing growth is modified in various ways. Among paired glands the removal of one causes increased size of the other. Here growth is compensatory. On the other hand Samuel found that the renewal of the feathers in a pigeon's wing was modified by disturbances of the circulation, and also that when the disturbance was practised on one side only it caused an arrest of growth in the feathers on both sides,

[1] Bizzozero, *Wien. Med. Blätt.*, 1894.

thus bringing to light a mechanism for symmetrical growth.[1]

The relations between geographical distribution and the colour and size of various birds and mammals of North America have been studied by Baird, Allen, and others.[2] They found that the largest individuals were obtained from localities at which the type was most numerously represented. Such localities are designated as the centres of distribution, and are assumed to be those in which that type finds the best conditions for existence. Many mammals and birds are found to have their centres of distribution in the northern regions, and so to diminish in size from the northern to the southern latitudes, thus showing that mere increase in temperature, which at first sight might be expected to favour size, does not do so in these cases. Reversed instances, where the centres of distribution are in southern latitudes, also occur, but they are not so numerous. Familiar are the enormous possibilities of variations in size among animals under domestication, and our small and large dogs and horses are sufficient indication of what can be done in part by selection. But it is quite in accordance with these facts to look upon size as a character mainly predetermined in the egg, and only to a moderate degree influenced by conditions acting through the lifetime of the individual.

The increase in stature and weight of the higher animals has been found to show a periodic variation in the rate at which it goes on, whether the increase is due to the multiplication of cells, or simply to their enlargement. In many eggs when dividing there are between the periods of actual division long intervals in

[1] Samuel, *Virchow's Archiv.*, 1887.

[2] Allen, *Radical Review*, 1890.

which the contents of the egg are apparently inactive. Cell multiplication and enlargement therefore appear to be distinctly rhythmical processes, but it must be understood that the cell, though seemingly quiet, is, in a chemical sense, constantly active, and attention is here directed only to the fact that the evident processes in cell multiplication and enlargement are periodic ; like similar phenomena this periodicity is particularly accentuated at some times and in some species, though it probably holds for all stages of growth in all animals, man included.

On bringing the foregoing facts together we find that the animal body is composed of cells, which are the lineal descendants of the fertilised ovum. Any active cell, having reached a point at which it no longer divides, grows larger, the cytoplasm increasing more rapidly than the nucleus. Variations in structure appear, the shape alters, and accompanying all these is a steady change in chemical constitution. Chemically, therefore, the cells are being continuously modified, even though as structural elements they may have an existence conterminous with that of the entire organism. The complex animal when formed is composed of different groups of cells, which, though they are fed by the same nutrient lymph, become structurally modified in such a manner that they are suited to the performance of limited and dissimilar functions, thus placing the several systems of tissues in a relation of mutual dependence. The size of the adult animal depends on the number and volume of the constituent cells. In development the determination of number precedes that of size, but the conditions controlling both processes are mainly resident in the cells themselves. The changes, however, which lead up to the final form of the body do not occur continuously, but are periodic, and the rate at which the growth processes go on diminishes from the first. With their cessation begins the period of decay.

Starting with these ideas as introductory to the study of growth, a few words may be devoted to an outline of the manner in which it is proposed to apply and expand them. The arrangement is determined by the wish to focus on the question of brain growth as many groups of facts as will lead to the better comprehension of it. The growth of the entire body is therefore first to be studied, and then, since the different tissue systems are mutually interdependent, such facts as bear on them separately will be introduced. In the skeleton it is the growth of the skull which is important for our purpose, since it exercises a direct influence upon the enclosed brain. We naturally pass from this to the growth of the brain itself, and the conditions which more directly modify it. Of course with it the spinal cord must be considered, as together they form a unit, the Central Nervous System. The normal brain first claims attention, and we shall begin by discussing from the statistical point of view its average weight and size; but since the brain is the main object in this study, it seems justifiable also to introduce some of the facts concerning its abnormal development. This will enable us to estimate in a measure the value of those speculations which involve the assumption that the delinquent and dependent classes in modern society are largely the victims of their own nervous organisation. The facts, however, up to this point have been gathered entirely from observations made on the organ taken as a whole, and their interpretation will become much more evident after considering the meaning of the size, number, and general arrangement of the elements of which the central nervous system is composed, for the nerve cell stands in the same relation to the system of which it is an element as the individual man does to the social organism. The architectural arrangement of these cells is orderly, and must be realised, both for the better understanding of the relations existing between the different portions of the nervous system and the body which it controls, as well as that the important fact of the localisation of function in different portions of the system may be intelligently discussed. Having presented the available facts under this head, the study of the different functions of the several portions of this system, or the study of localisation in the nerve centres, will be undertaken. This completes the anatomical side of the subject, and the remaining chapters deal mainly with physiological problems.

It is desirable at the start to understand the manner in which the nerve cells react, and the laws controlling the habits and rhythms, of which they are capable. Of many of these processes we are conscious, but, fortunately for us, there are also many of which we are quite uninformed. Naturally all reactions induce fatigue, followed by repose, which permits recuperation. In this rhythm of fatigue and recovery there are wide individual variations, and in considering these variations we get a glimpse of temperament. Finally there is a point in the history of the individual when, both mentally and physically, the capacity to feel and to act diminishes. The failure is not necessarily equivalent in all directions, but there are certain general features in these changes which it is important to describe. With this foundation of anatomy and physiology the significance of training and educational methods may be properly considered, and as a division of this general topic it may not be amiss to add a word concerning sex in education. Finally the attempt will be made to summarise the facts most important for the best understanding of the form and function of the human central nervous system, and its changes during growth, presenting the wider view, and showing the more general bearings of the observations which have been discussed.

Note to p. 37.—In cases of "sporadic cretinism"—arrested bodily and mental development, associated with the atrophy of the thyroid gland—thyroid-feeding has produced in some instances a remarkable increase in stature together with corresponding mental improvement. Under such treatment a boy of nine years grew four and one quarter inches in height in eight months, and at the same time showed mental improvement. Similarly another boy of nineteen years grew four and one half inches within a year. The thyroid-feeding appears to supply some substance needed for growth, but lacking in these individuals. *Vide* Thomson, *Edin. Med. Jour.*, May, 1893; Shuttleworth, *The Hospital*, March 30, 1895.

CHAPTER II.

INCREASE IN THE WEIGHT OF THE ENTIRE BODY.

Terms—Divisions of the life cycle—Growth and development —Growth of the entire body—Variations in the specific gravity of the different systems—Proportion of water—Weight the best measure of growth—Change in the proportions of the body with age—Data on weight—Corrections for clothing—Descriptions of curves—Absolute increase in weight—Composition of statistics—General and individual methods—Effects of illness—Influence of environment—Influence of size attained on subsequent growth—Weight and stature of first-born —Growth before birth—Determination of weight at birth—Increase during the first year—Rate of growth—Unit of energy.

UNDER the term growth are included several sets of changes which should be distinguished. In following a mammal from birth to the completion of its individual existence, a constant series of alterations in its physical characters is to be noted. From infancy to maturity the animal increases in all diameters and in weight. To these physical changes the term growth has commonly been limited. But after reaching maturity there is a period when the changes become very slow, and this is later followed by the involutionary processes of old age. The second and third acts are as important a part of the life cycle as the first, and yet in the last the physical changes are predominantly those of decrease. The possibilities of confusion do not, however, end here.

Both the size and weight of the animal may vary by alterations in the amount of fat, and yet such variations do not, strictly speaking, belong to the phenomena of growth, any more than does the increase in stature which occurs each day during the hours of sleep. Moreover, to the term growth, implying increase in size, there is usually attached the notion of an increase in power and capacity for work which are qualitative changes. The word growth, therefore, has certain disadvantages, by virtue of its several connotations, so that it will be better to employ, so far as possible, other terms intended to be more exact.

Between the first changes in the fertilised ovum, and the somatic death of the individual, is comprised a life cycle. Within this cycle we recognise the following subdivisions: the embryonic period, when the organs are being formed; the fœtal period, between the formation of organs and the time of birth; infancy, the period of dependence upon the mother—in medical jurisprudence extending to the time when the milk teeth begin to be shed—childhood, from the beginning of independence to the age of puberty; youth, from puberty to the completion of the increase in both stature and weight; maturity, from the completion of growth to the onset of uncompensated decay; old age, from the beginning of uncompensated decay to death. Much stress must not be placed upon any such division, since it is only an attempt to break up a really continuous process, an attempt which can never be satisfactory, but which has its apology in convenience, and in the fact that it defines some useful terms. Within the life cycle and its periods cells undergo first increase, then decrease in size, weight, and number. These may be designated as growth changes. At the same time adaptive alterations take place in cell structure and constitution, and these are the changes of development. The term development, then, will be used when changes in the arrangements or chemical constitution of the cell substances are implied, changes which are qualitative, as contrasted with those evidently quantitative. After the cells in the nervous system have developed in some measure, they enter into relations with one another, which are closer than those between the undeveloped cells. The establishment of these relations constitutes the organisation of the system—a change by which not only the strength, but also the complexity and precision of its reactions are all increased. As a matter of fact, both qualitative and quantitative modifications go hand in hand, so that the distinction thus made is merely formal. It will be possible to study the growth of the central nervous system with a greater advantage if the principal facts relating to the growth of the entire body are first passed in review, and to this task we now turn.

The elements of the human body while growing are continually undergoing chemical variations, which in turn tend to cause variations in their specific gravity. It is, therefore, to be remarked, that so far as the change in the weight of man is concerned, variation in chemical composition plays but an unimportant *rôle*. Taking the body as a whole, the tissues of infants possess the greater, those of the aged the smaller percentage of water. For the fœtus of one month Fehling[1] gives the percentage of water as 97·5 per cent. At birth this falls to 74·7 per cent. Bischoff found it to be 66·4 per cent. at birth, whereas in an adult it was only 58·5 per cent. of the weight of the body. The discrepancy between these two determinations of water at birth is merely an expression of the wide variations in this relation. When, however, the separate organs of the new-born and aged are compared in detail, it is by no means the case that in each instance the percentage of water in the new-

[1] Fehling, *Arch. f. Gynækol.*, 1877.

born is the greater ; the total difference is therefore the algebraic sum of two opposite tendencies, causing a loss of water from some of the tissues, and an increase of it in others. From the comparative constancy of the specific

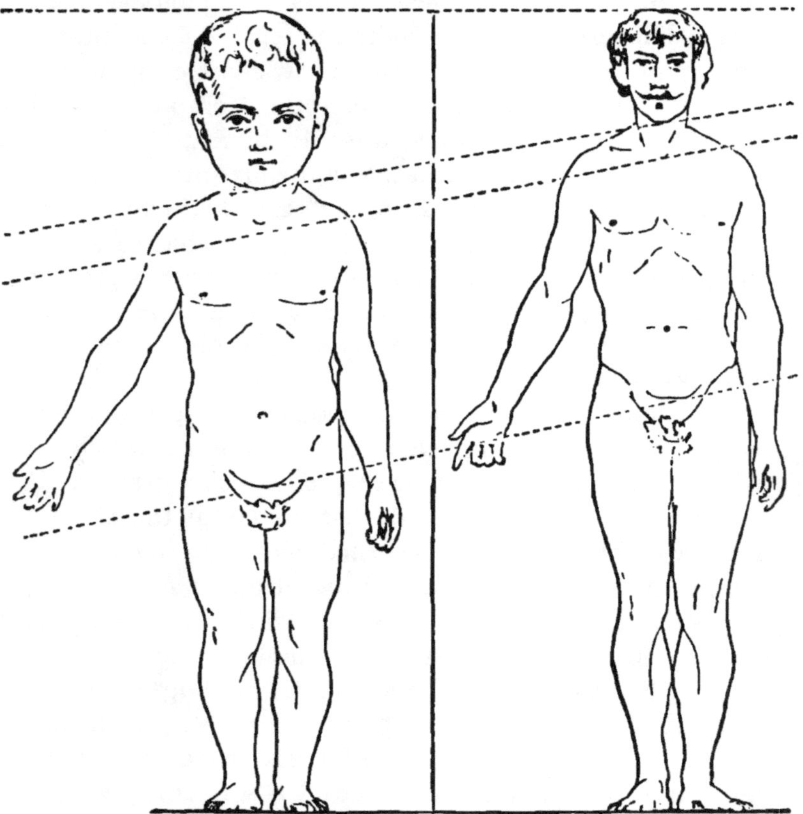

FIG. 8.—Showing the relative proportions of the child and adult. (Langer.)

gravity after the first years of life it follows that in measuring the increase of the entire body, the increase in weight and in volume will be very nearly the same.

The study of growth by weighing furnishes the best measure of the total changes of the body. It is easy to see why this measure should be best, if we consider for a moment the significance of linear measurements which are sometimes taken in place of it. Our bodies are developed in three dimensions, and give an irregular geometrical form which, moreover, changes with age. In consequence of this, the variation in any linear measure, as stature, for instance, serves only as a rough index of the changes which have taken place in the other two diameters. The accompanying figure, taken from Langer,[1] shows at a glance that a man is not a big child, for when the child is enlarged to the size of a man, as in this illustration, its short legs, large trunk, and enormous head are in sharp contrast with the same parts in the adult.

This change in geometrical form is well exemplified by a citation from Thoma.[2] Starting with the figures of Quetelet, he makes the following tabulation :—

Male—new-born,	stature 500 mm.,	body-weight 3·1 kilos.
Man at 30 years	„ 1686 „	„ 66·1 „

If we take now the specific gravity of the entire body of the new-born as ·90,[3] and that of the man at thirty years as ·93, we get as corresponding volumes—

Male, new-born	3·44 lit.
Man, 30 years	71·08 „

By referring back to the first table we see that during the first thirty years of life stature has increased

[1] Langer, *Anatomie d. äusseren Formen d. mensch. Körpers*, 1884.

[2] Thoma, *Grösse und Gewicht der anatomischen Bestandtheile des menschlichen. Körpers*, 1882.

[3] As the specific gravity of almost all the solid and fluid constituents of the body is more than 1, it is at first sight curious that the entire body should have a specific gravity below 1. These observations were, however, made on intact bodies, the air and gases remaining in the cavities. Hence the lower figures which were obtained. As the relative specific gravity is here the point of interest, the peculiarity of the absolute figures need not be further discussed.

1686 ÷ 500 = 3·37 fold, and the volume of the body 71·08 ÷ 3·44 = 20·66 fold. If during this period of thirty years the geometrical relations at birth had persisted, the increase in volume of the body would have been proportional to the cube of the increase in stature, but the cube of 3·37 is 38·27, which is almost twice as much as the increase in volume which has been observed. It is thus shown that the geometric relations existing at birth must change during the progress of growth. For this reason the study of stature, important as it is, does not furnish data from which the changes in the entire body can be inferred except in a most general way.

In dealing with the data relating to change in weight, some care must be exercised in selection of the observations to be employed as standards. Of course, since the different races of mankind differ widely in their final weight, and probably in the details of the manner in which they attain it, the individuals who are to be tested by reference to a chosen standard, should be compared only with data derived from the study of those belonging to the same race and growing up under conditions as nearly similar as it is possible to have them. There have been already gathered in Denmark, Germany, Sweden, England, Russia, and the United States, a very large number of observations on the weights of school-children at different ages. These observations are unfortunately limited in range to the school ages, viz., from about four to twenty years, and thus leave out the important years before four and after twenty, the record for which it is desirable to have in order to compare, from birth to maturity, the growth of the central nervous system with that of the body. For the following discussion it seems best to employ but a single series of observations, and for this purpose those on the general population of Great Britain published by Roberts have been selected.[1] These observations have been chosen in preference to some others because of their range, and their special applicability to English-speaking people. They are those employed by Galton in his Life History Album. Table 2 gives the figures for the average weight of persons between birth and twenty-four years.

TABLE 2.—GIVING THE AVERAGE WEIGHT OF ALL CLASSES OF THE POPULATION IN ENGLAND AT YEARLY INTERVALS. THE RECORD IS IN POUNDS AVOIRDUPOIS. THE WEIGHT IS TAKEN WITH INDOOR CLOTHING. (*From Roberts.*)

MALES. General Population—All Classes. Town and Country.			FEMALES. General Population—All Classes. Town and Country.	
Age last Birthday.	No. of Observations.	Average weight in lbs.	Average weight in lbs.	No. of Observations.
Birth.	451	7·1	6·9	466
1	—	(24·0)	20·1	8
2	2	32·5	25·3	9
3	41	34·0	31·6	30
4	102	37·3	36·1	97
5	193	40·0	39·2	160
6	224	44·4	42·0	178
7	246	50·0	47·5	148
8	820	55·0	52·1	330
9	1425	60·4	55·5	535
10	1464	67·5	62·0	495
11	1599	72·0	68·1	456
12	1786	77·0	76·4	419
13	2443	83·0	87·2	209
14	2952	92·0	97·0	229
15	3118	103·0	106·3	187
16	2235	119·0	113·1	128
17	2496	131·0	115·5	74
18	2150	137·4	121·1	64
19	1438	140·0	124·0	97
20	851	143·3	123·4	128
21	738	145·2	122·0	59
22	542	148·0	123·4	53
23	551	148·0	124·1	29
24	483	148·0	121·0	19

[1] Roberts, *Manual of Anthropometry*, 1878.

The accompanying Fig. 9 expresses the facts of the table in the form of curves.

In order to correct the weight for clothing, the observations on the proportional weight of clothing made by Bowditch may be employed.[1]

TABLE 3.—WEIGHT OF INDOOR CLOTHING GIVEN IN PERCENTAGE OF THE BODY-WEIGHT. BOSTON SCHOOL-CHILDREN. (*Bowditch.*)

AGE IN YEARS.	MALES.	FEMALES.
5—8	6·5—7·2	6·5—7·5
9—12	7·9—9·9	6·8—6·9
13—15	7·8—8·4	5·8—7·3

If the construction of the chart is understood, it will be seen that the records of increase, both in weight and stature, show that the absolute increase from year to year is not the same; for were it so, the lines indicating increase would be straight. As a matter of fact, they are sinuous, thus indicating variations in rate. The greater the annual increase, the more nearly the record approaches the vertical, and the slower it is, the more horizontal becomes the curve. Glancing at the curve for the males, it is seen that for the first year of life increase in weight is rapid. This is followed by a period of slower increase up to seven years of age. From here the weight increases again more rapidly up to sixteen years, with a maximum rate between sixteen and seventeen years. The rate then falls to twenty-five years, and from that time on, if the curve were continued up to fifty years, the increase would be slight, although continuous. In this later period, from twenty-five years onwards, the accumulation of body-fat is the chief cause of the

[1] Bowditch, *The Growth of Children.* Report of the State Board of Health of Massachusetts, Boston, 1877.

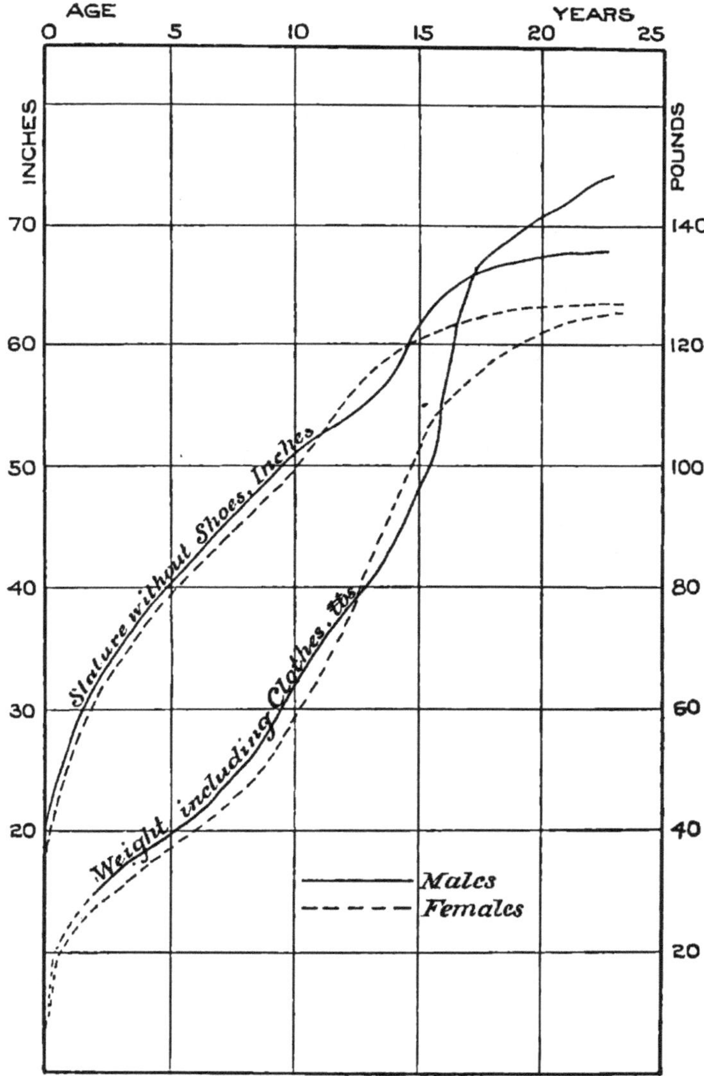

FIG. 9.—Showing curves of increase in weight and stature for both sexes. Curves based on the figures in Table 2. (Roberts.)

increase in weight just noted. On comparing the curve for the females with that for the males, by far the most striking feature is the similarity of the two. This is the more to be emphasised as in the further discussion it is the points of difference which will be specially described. The curve for the weight of the females follows that for the males very closely up to the sixteenth year; at that time there commences a divergence which is most important, and which leaves the females during the remaining years of life some 25 lbs. or more behind the males. This divergence is brought about by a period of more vigorous growth in the males between the fifteenth and eighteenth years.

But before reaching this point of final separation, there are in these curves at least three features of great interest. Both curves start at birth from almost the same point, that for the male, however, being slightly higher, thus indicating that at birth the male infant is heavier. Up to the end of the first year the curves diverge rapidly, and since the steeper curve indicates the greater absolute increment, we see at once that the male has grown more during this time. From this point to the tenth year the two curves diverge much more slowly, running at times quite parallel to one another. About the eleventh year the curve for the female becomes more perpendicular. The curve for the male still continues at the same angle as before, and the result is, that at the twelfth year, or a little later, the curves converge, and, being close together, that for the females crosses that for the males. From this point to the sixteenth year the girls are heavier than the boys. This second period is closed, as the curve for the female becomes more horizontal and that for the male more perpendicular, so that they again cross. Beyond this is the wide divergence already described. In both the males and females there are, then, two principal periods of acceleration in weight increase. The first is observed in both sexes in the years immediately following birth, and is most marked during the first year. At the tenth year weight increase in the female is accelerated for the second time, and later the curve for the female crosses that for the male. As indicated on these curves this acceleration lasts for about four years. The acceleration in the male does not begin until a year or two later, but is greater than in the female, and lasts for a longer time. It therefore appears that the female exhibits the second acceleration somewhat earlier in life, and for a shorter time; and by reason of this there occurs the curious convergence of the two curves, a feature which all careful statistics on the growth of children show.

The relations just described are illustrated in another way by the curves of Stephenson given in Fig. 10,[1] curves which are based on the observations by Bowditch as well as those by Roberts. They show the annual absolute increase in weight for the two sexes.

Returning to the statistics by Roberts, it becomes evident that the information to be derived from them fails somewhat in explicitness because they have been formed by weighing and measuring many children at different ages, and not following the growth of individuals through the entire period of years. While in a large way the "general" and "individual" methods, as they have been called respectively, give similar results, yet there are some details in which they necessarily differ. Different children do not undergo the pre-pubertal acceleration in growth between the same ages, and while in the case of boys, as shown by the general method, it may begin even before the eleventh year, and

[1] Stephenson, *Lancet*, 1888.

extend beyond the seventeenth, the period in any individual is usually much shorter. Thus any such growth change which is common to all individuals, yet variable

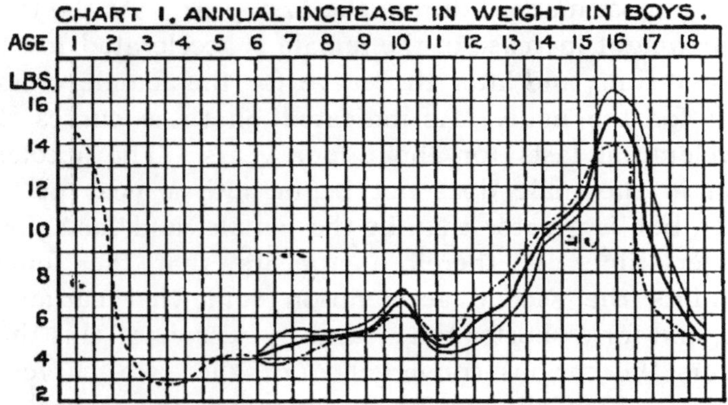

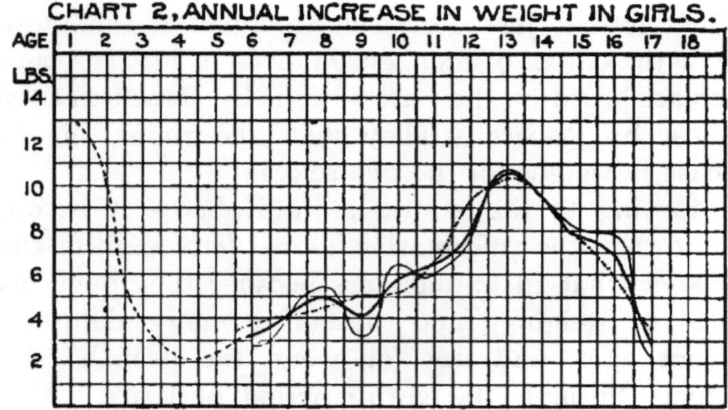

FIG. 10.—Charts 1 and 2, showing annual absolute increase in weight in boys and girls. Based on observations by Bowditch ————, and those by Roberts —·—·—·. The heavy line ———— gives the average of these two. Before the sixth year the record is from other sources. (Stephenson.)

in the years between which it occurs, must always have an undue extension over a curve of growth derived by

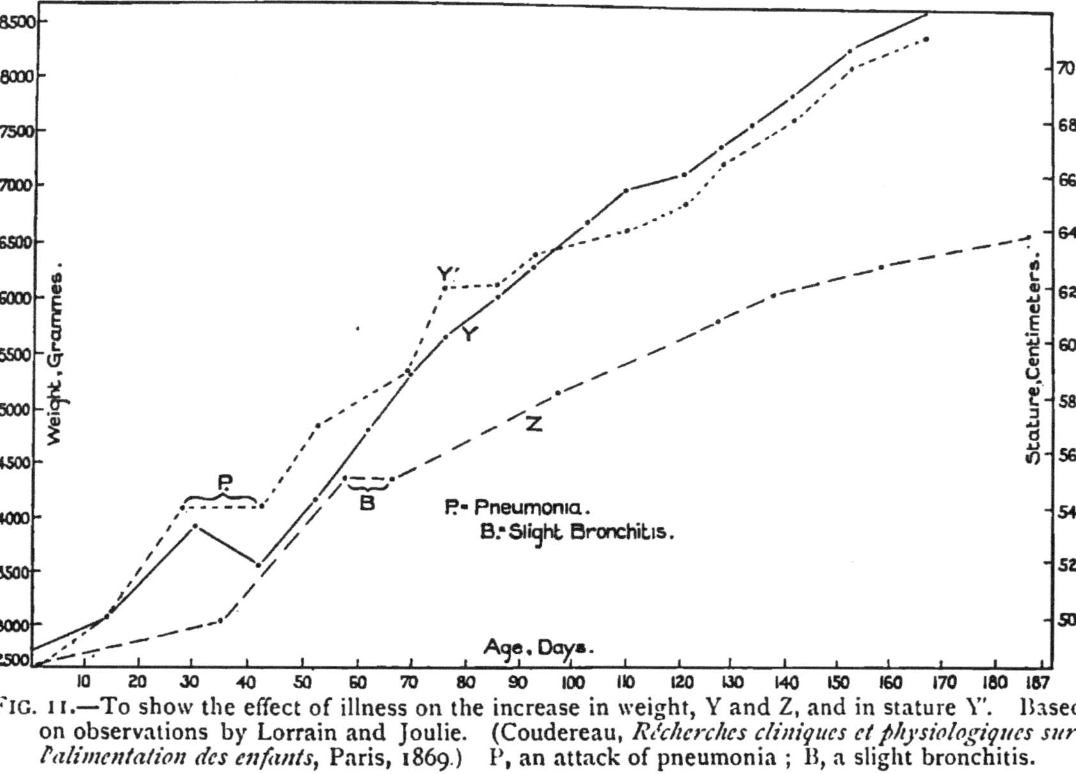

FIG. 11.—To show the effect of illness on the increase in weight, Y and Z, and in stature Y'. Based on observations by Lorrain and Joulie. (Coudereau, *Récherches cliniques et physiologiques sur l'alimentation des enfants*, Paris, 1869.) P, an attack of pneumonia; B, a slight bronchitis.

the general method. It is also perfectly clear that the number of observations on which the first part of the curve is based, is too small; and that the course of the curve for the first five years is not nearly so well determined as for the next twenty. So, too, the different classes in the community are not proportionately represented at the different ages, and another slight error is thus introduced, but on the principal points these data are sufficiently accurate.

The differences in individual growth are dependent on a variety of other conditions. It has often been observed that illness retards the growth of children, and that on recovery from illness there is a tendency to recover the weight and stature which should normally have been attained; hence just after illness the child may grow with unusual rapidity. These facts are admirably illustrated by Fig. 11, taken from K. Vierordt, where in one case Y and Y′, pneumonia causes the increase in stature to cease and weight to be lost, and in Z, a slight bronchitis prevents for a short time any gain in weight.[1]

The younger the child the more readily the recuperation is accomplished. Although external conditions have but a slight influence on the general course of the growth curve, yet the specially favoured classes in England and the United States are plainly above the average of the community in both weight and stature. At the same time it is hardly proper to assume that

[1] The measurements in this figure are in grammes and centimeters. For those not familiar with the metric system, the following equivalents are given :—

=1 Inch

=3 Centimeters

One inch is approximately 2·5 cm.
One pound avoirdupois, 453·59 grms.
One ounce „ 28·35 „

surrounding conditions are here the sole factor determining the difference. The Indians of North America living on the mountains are smaller than those living on the plains, a correlation which may be associated with the greater abundance of food on the plains.[1] But under like conditions half-breeds tend to be taller than their parents, and have a slightly different curve of growth,[2] here again indicating that much must be allowed to the peculiarities of the species on which any set of external conditions happens to be acting. Further, Boas has found that first-born children are, in the end, taller and heavier than others of the same family.[3] We are aware of no conditions compatible with life in which the general character of the growth curve with its acceleration during adolescence can be altered. Minor variations may, however, arise. From a study of the relations of the curves for the two sexes under varying conditions, it appears (Roberts' Tables) that the environment has more influence on the male than on the female, and that under the most favourable conditions the two curves run more nearly parallel owing to the better growth of the males; but the cause of this may be sociological rather than strictly biological.

The weight of the individual at any period of development must be looked upon as one factor influencing further growth. Möhring found that the loss of weight[4] occurring in infants during the first few days of life was more marked in those whose weight at birth was small. Bowditch,[5] and also Porter,[6] have

[1] Boas, *Mem. of the Internat. Congress of Anthropology*, Chicago, 1893.

[2] Boas, *Pop. Sci. Monthly*, 1894. [3] Boas, *Science*, March 1, 1895.

[4] Möhring, *Inaug. Diss*, Heidelberg, 1891.

[5] Bowditch, *Growth of Children.* Twenty-second Annual Report of the State Board of Health of Massachusetts, Boston, 1891.

[6] Porter, *Transaction Academy of Sciences*, St. Louis, 1894.

sought to determine in school-children the significance for subsequent growth of the size already attained. The determination is, however, extremely difficult, since the small children are not necessarily transformed into small adults, nor is the small child at one age necessarily the small child at the next period in its growth.[1]

While the statistics on weight just presented may be all that are needful for our present purposes, it would also be of value to know the course of weight increase before birth. The table here given is compiled from observations gathered by Vierordt.[2]

TABLE 4.—COMPILED FROM VIERORDT TO SHOW THE INCREASE IN WEIGHT FROM THE OVUM TO THE INFANT AT BIRTH.

AGE BY WEEKS.	WEIGHT IN GRAMMES.	WEIGHT IN GRAMMES.	AGE BY WEEKS.
0 (ovum)	·0006	635	24
4	—	1220	28
8	4	1700	32
12	20	2240	36
16	120	3250	40 (Birth)
20	285		

This shows amply how enormous is the increase in weight during the first three-quarters of a year. From ·0006 grms. to 3250 grms. is an increase of more than five million fold, so that compared with this later growth is a very feeble performance. The weight of the child at birth is the resultant of several modifying conditions. Minot [3] has brought together the data on this point, and his facts are here summarised. The size of the child is correlated with the size of the mother. Gassner states that the average weight of the child is 5·23 per cent. of the maternal weight. It is also found that the heaviest children are born to mothers aged about thirty-five years.

[1] Boas, *Science*, March 1, 1895.
[2] H. Vierordt, *Daten und Tabellen*, 1893.
[3] Minot, *Human Embryology*, 1892.

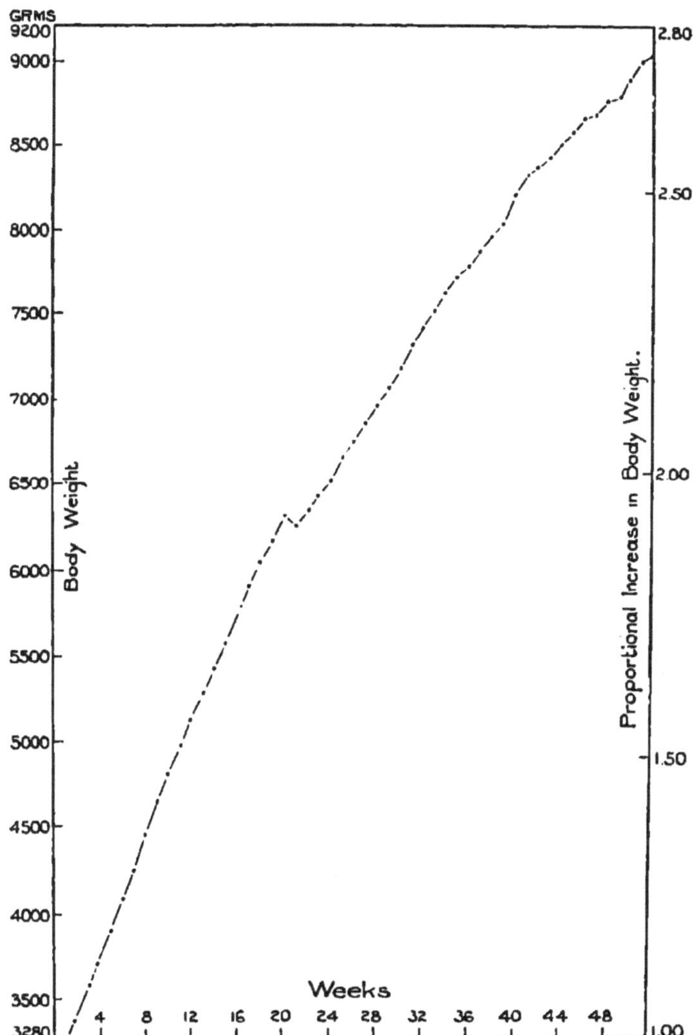

FIG. 12.—Composite Table, showing the increase of the infant in weight during the first year. Calculation by Dr. Meeh. (K. Vierordt.)

This may be a result of age alone or may be complicated by the fact that previous pregnancies tend to increase the weight of the last born child. To these conditions are also to be added the length of gestation and individual differences in the growing power of the fœtus, on both of which the foregoing conditions must exert an influence.

A record for growth during the first year of life is given from Vierordt after Meeh.[1] This shows the rapid increase in weight, which, with the exception of an insignificant regression during the twenty-first week, is continuous. (Fig. 12.)

The rate of growth has already been mentioned, and it will be important to determine for man the rate of the increase in weight. The facts at first sight do not arrest the attention, but when we stop to consider, it is plain that there must be both variations, and also a great diminution in the rate, similar to that pointed out by Minot in his study of guinea-pigs. If the weight of the body at any date be taken as a basis, the percentage increase on this figure during the next unit of time can be calculated, and so on for any number of successive intervals. Treating in this manner the figures for the increase in weight, as given by Roberts, they yield a curve as shown in Fig. 13, where from birth to the first year there has been an increase in weight of 240 per cent. for the males, and 190 per cent. in the females. But this rate diminishes during the succeeding years with astonishing rapidity.

This curve indicates that the weight increase which we measure between birth and maturity may be regarded as the latter end of a dwindling capability, in which there is a slight revival at the age of puberty, a revival representing the last strong impulse in the direction of enlargement.

[1] K. Vierordt, *Anatomie und Physiologie des Kindesalters*, 1881.

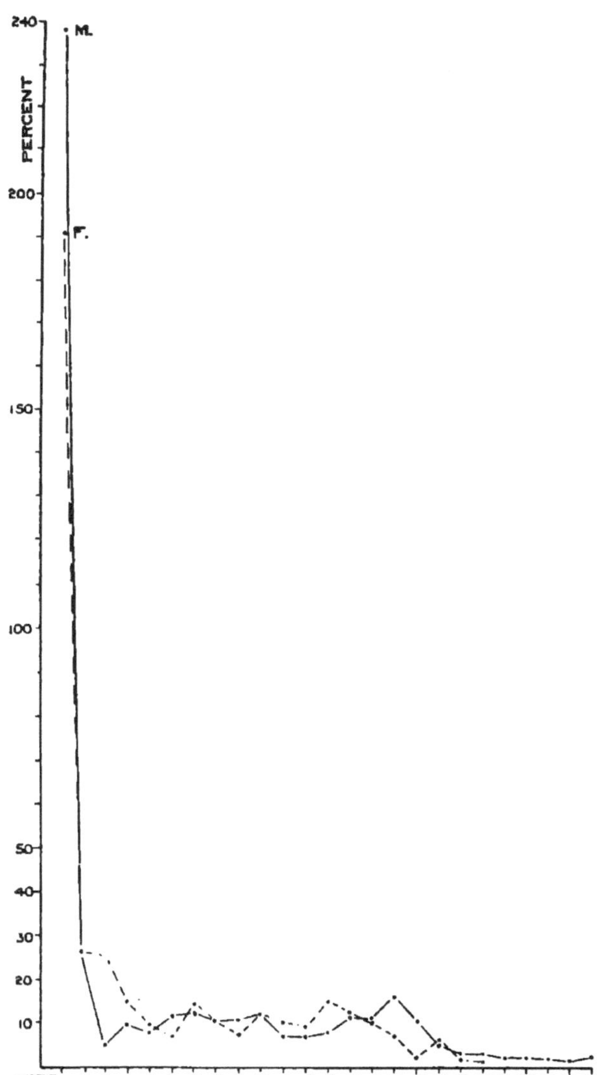

FIG. 13.—Curve to show the rate of increase in weight during the first twenty-five years of life. M. males; F. females. The percentages of increase are measured along the vertical axis, the ages along the horizontal one. (Based on Roberts, Table 2.)

The mystery of this running down of the animal machine awaits explanation, though of course the mechanical hindrances to nutrition following increase in bulk, and the greater exertion necessary to obtain food after separation from the mother, have long been recognised. There is, however, something beyond mechanics in the problem. Indeed there is some ground for considering each animal as possessed of a certain limited store of potential energy with which to go through its life, but the complexity of the manner in which this energy may be expended is so great, and the combinations of its expression so manifold, that as yet it is possible only in the most general way to guess at the capacity of a given individual. We admit that disease is fundamentally exhausting, that excessive exertion and privation leave an indelible mark on the bodily organism, and yet there is so much art in the manner of putting forth exertion, that at present we are quite without means to adequately record its expenditure, or to explain how structural deficiencies may be masked, for the measure of the whole man is neither the number of pounds that he can lift, nor facts that he can discover, nor the influence that he can exert upon his fellows, nor yet the age to which he can attain, but something of all these and what they stand for, taken together.

CHAPTER III.

WEIGHT INCREASE OF DIFFERENT PARTS OF THE BODY AND INCREASE IN STATURE.

Tissue systems—Weight changes—Significance of changes in proportion—Weights of organs—Number and size of elements—Hypertrophy and hyperplasia—Statistical difficulties—The dead form a series different from the living—Premaxima—Deviation from the mean—Mortality curve—Relation to the statistics of weight—Weight and longevity—Growth an expression of vigour—Stature—Direction of first growth—Rate of increase—Rhythm of growth—Relations of the divisions of the day and year to growth.

THE observations contained in the previous chapter relate to the entire body, while here I shall introduce facts bearing on some of its anatomical subdivisions. For the purpose of this further study the entire body may be resolved into the group of tissue systems which compose it. Indeed it is almost necessary to approach the subject in this way, because the different tissues, of which the nervous system is one example, are so mutually dependent upon one another for the manner in which they develop, that important modifying conditions are overlooked, when any one system is considered alone.

Histologists recognise in the body a series of structural elements, designated as epithelial, connective, nervous, and muscular. From these, combined in varying proportions, are built up the different organs of the body, which may again be ranged in larger groups or systems. The bony skeleton is connective tissue, rendered rigid by a deposit of lime-salts. The system of the voluntary or skeletal muscles is predominantly composed of muscular tissue ; while in the case of the nervous system this dominance of characteristic tissue is even more marked.

Within the thoracic and abdominal cavities of the body are the several viscera, and these last may be removed and studied separately. What remains after their removal is the skeleton enclosed by the muscles, and in turn enclosing the central nervous system, the whole mass being covered by the skin. By proper methods these several portions may be separated from one another. The facts obtained by such a separation are in part shown in the following table on the weight and composition of the human body, as given by Bischoff[1] :—

TABLE 5.—SHOWING THE ABSOLUTE WEIGHT OF SIX PERSONS OF DIFFERENT AGES (FOUR MALES, TWO FEMALES), AND THE PERCENTAGE OF THE TOTAL BODY-WEIGHT REPRESENTED BY THE PARTS NAMED. (*E. Bischoff.*)

	PREMATURELY BORN. MALE.	NEW-BORN		YOUTH.	WOMAN.	MAN.	
		FEMALE.	MALE.				
Weight of entire body in grammes	364	2969	2300	35547	55400	69668	
Age, years ...	Fœtus 21 wks.	0	0	16	22	33	
Skeleton	20·3	15·7	17·7	15·6	15·1	15·9	Percentage of Body-Weight.
Muscles	22·3	23·9	22·9	44·2	35·8	41·8	
Thoracic viscera ...	2·7	4·5	3·0	3·2	2·4	1·7	
Abdominal viscera.	12·3	12·1	11·5	12·6	8·2	7·2	
Fat	14·8	13·5	20·0	13·9	28·2	18·2	
Skin	—	11·3	—	6·2	5·7	6·9	
Brain	18·5	12·2	15·8	3·9	2·1	1·9	

[1] E. Bischoff, *Zeit. f. rationelle Med.*, 1863.

In this table let us first consider the figures relating to the male child and to the man. On comparing the proportion of the different systems in the new-born with the adult, the following relations are worthy of remark: the percentage of the skeleton, fat and skin, taken together, is but slightly smaller in the new-born. The percentage of the viscera in the new-born is nearly twice, that of the central nervous system more than eight times, that found in the adult, whereas the proportional weight of the muscles shows only a trifle more than one-half its adult value. The purely constructional parts of the body, the skeleton, fat and skin, which are also formed predominantly of connective tissue, have therefore not varied their proportion during growth; while the nutritive and controlling system, that is, the abdominal and thoracic viscera and the brain, have undergone a relative diminution, having in a most remarkable way been outgrown by the muscular system.

On comparing the female child with the woman, most of the preceding statements might be repeated. On comparing the two sexes it is found that the increase in the proportion of the muscles is less marked in the woman than in the man, while the increase in the fat is greater. Indeed in both the female child and the woman the percentage of fat is high as compared with the other sex.

On considering the new-born child, we are impressed by the fact that the great vegetative system represented by the viscera of the trunk is in early life of proportionally large bulk. Since with advancing age the constructive processes become slower and finally cease, the duty of maintenance being alone performed, it may be inferred that the demands on this system made by the enlarging muscles soon match its powers of performance. With a slight change of terms, the relations of the central nervous system to the body at large may be similarly described. That a normal relation between growing parts and the central system is necessary to their mutual development is certain; but whether through the nerves the part under control directly receives anything more than the stimuli which keep it in a state of healthful excitement, and so enable it to make the best use of the surrounding lymph, is an open question. However this may be, the relative bulk of the central nervous system has strikingly diminished in the adult as compared with the new-born.

As bearing on the proportional development of the body, the observations of Bischoff have been quoted, and though at present they represent the best part of our meagre knowledge of this subject, yet in the future we must hope to learn something of the effect upon the various systems of the conditions which are found to influence the growth of the body as a whole. All the older data have also been collected by Vierordt, and to his various tables the reader is referred.[1] The foregoing table has also made plain that the several tissues and organs increase in weight in very different proportions. Among these, the viscera of the trunk make but a poor showing, being proportionately much diminished in the adult; yet we have rather extensive data on the change in the weight of the several viscera, and these are worth examination.

The facts bearing on this line of investigation have been put by Vierordt[2] in the form of tables, which are given below. Though the organs involved amount at birth to only about 20 per cent. of the weight of the entire body, more than half of which is represented by the brain, and though they all belong to

[1] H. Vierordt, *Daten und Tabellen*, 1893.
[2] H. Vierordt, *Arch. f. Anat. and Physiol.*, 1890.

that group, the relative weight of which decreases with age, nevertheless the variations observed in them are significant.

TABLE 6.[1]—SHOWING THE ABSOLUTE INCREASE, PERCENTAGE INCREASE, AND PERCENTAGE OF THE BODY-WEIGHT, OF THE PRINCIPAL VISCERA OF MAN BETWEEN BIRTH AND THE AGE OF TWENTY-FIVE YEARS—MALES. (*H. Vierordt.*)

MALES.

(1) *Weight of the several organs in grammes.*

	AGE.	BRAIN.	HEART.	RIGHT LUNG.	LEFT LUNG.	LIVER.	KIDNEYS.	SPLEEN.	BODY-WEIGHT IN KILOS
Absolute increase.	New-born	381	24	30	24	142	23	11	3·1
	1 year	945	41	83	74	333	73	20	9·0
	5 ,,	1263	81	130	111	539	115	57	15·9
	10 ,,	1408	128	236	251	837	161	88	25·2
	15 ,,	1490	199	383	368	1306	240	145	41·2
	20 ,,	1445	305	514	449	1561	296	186	59·5
	25 ,,	1431	301	513	482	1819	306	163	66·2

(2) *Amount of increase, weight at birth being taken as unity.*

	AGE.	BRAIN.	HEART.	RIGHT LUNG.	LEFT LUNG.	LIVER.	KIDNEYS.	SPLEEN.	BODY-WEIGHT IN KILOS
Relative increase.	New-born	1	1	1	1	1	1	1	1
	1 year	2·48	1·75	2·76	3·08	2·3	3·12	1·92	3
	5 ,,	3·32	3·43	4·35	4·63	3·8	4·92	5·40	5
	10 ,,	3·70	5·41	7·82	10·44	5·9	6·90	8·28	8
	15 ,,	3·91	8·45	12·67	15·38	9·2	10·29	13·68	13
	20 ,,	3·79	12·94	17·01	18·78	11·0	12·72	17·57	19
	25 ,,	3·76	12·74	16·97	20·14	12·8	13·12	15·38	21

(3) *Percentage value of the total body-weight at the ages given.*

	AGE.	BRAIN.	HEART.	RIGHT LUNG.	LEFT LUNG.	LIVER.	KIDNEYS.	SPLEEN.
Percentage of Body-Weight.	New-born	12·29	0·76	0·94	0·77	4·6	0·75	0·34
	1 year	10·50	0·46	0·92	0·82	3·7	0·81	0·23
	5 ,,	7·94	0·51	0·82	0·68	3·4	0·72	0·36
	10 ,,	5·59	0·51	0·94	0·99	3·3	0·64	0·35
	15 ,,	3·62	0·48	0·93	0·89	3·2	0·58	0·35
	20 ,,	2·43	0·51	0·86	0·75	2·6	0·50	0·31
	25 ,,	2·16	0·46	0·77	0·73	2·8	0·46	0·25

[1] Tables 6, 7 have been modified by omitting decimals. Professor Vierordt has informed me that in his paper of 1890 an error was made in quoting the figures of Quetelet for the body-weight. These tables here given contain the proper figures for body-weight, and have otherwise been corrected in so far as this change made revision necessary.

TABLE 7.—SHOWING THE ABSOLUTE INCREASE, PERCENTAGE INCREASE, AND PERCENTAGE OF THE BODY-WEIGHT OF THE PRINCIPAL VISCERA OF MAN BETWEEN BIRTH AND THE AGE OF TWENTY-FIVE YEARS—FEMALES. (*H. Vierordt.*)

FEMALES.

(1) *Weight of the several organs in grammes.*

	AGE.	BRAIN.	HEART.	RIGHT LUNG.	LEFT LUNG.	LIVER.	KIDNEYS.	SPLEEN.	BODY-WEIGHT IN KILOS.
Absolute increase.	New-born	384	24	32	23	164	23	11	3·0
	1 year	872	33	74	75	276	58	21	8·6
	5 ,,	1221	80	180	137	566	104	48	15·3
	10 ,,	**1284**	120	270	260	850	160	85	23·1
	15 ,,	1238	250	353	331	1420	235	122	40·0
	20 ,,	1228	243	438	365	1568	258	146	53·2
	25 ,,	1224	261	458	417	1664	291	173	54·8

(2) *Amount of increase, weight at birth being taken as unity.*

	AGE.	BRAIN.	HEART.	RIGHT LUNG.	LEFT LUNG.	LIVER.	KIDNEYS.	SPLEEN.	BODY-WEIGHT IN KILOS.
Relative increase.	New-born	1	1	1	1	1	1	1	1
	1 year	2·27	1·37	2·31	3·18	1·68	2·50	1·90	3
	5 ,,	3·18	3·35	5·64	5·85	3·45	4·50	4·44	5
	10 ,,	3·34	5·00	8·46	11·11	5·18	6·93	7·87	8
	15 ,,	3·22	10·42	11·08	14·14	8·68	10·13	11·27	13
	20 ,,	3·19	10·10	13·82	15·59	9·56	11·16	13·48	18
	25 ,,	3·19	10·86	14·36	17·82	10·12	12·61	16·04	18

(3) *Percentage value of the total body-weight at the ages given.*

	AGE.	BRAIN.	HEART.	RIGHT LUNG.	LEFT LUNG.	LIVER.	KIDNEYS.	SPLEEN.	BODY-WEIGHT IN KILOS.
Percentage of Body-Weight.	New-born	12·81	0·80	1·06	0·78	5·5	0·77	0·36	
	1 year	11·39	0·38	0·85	0·86	3·2	0·67	0·23	
	5 ,,	7·98	0·52	1·17	0·89	3·7	0·68	0·31	
	10 ,,	5·56	0·52	1·16	1·12	3·7	0·69	0·36	
	15 ,,	3·09	0·62	0·88	0·82	3·6	0·58	0·30	
	20 ,,	2·31	0·45	0·82	0·68	2·9	0·48	0·27	
	25 ,,	2·23	0·47	0·83	0·76	3·0	0·53	0·30	

Here are recorded the weight changes in the brain, heart, right lung, and left lung, kidneys (taken together), the spleen, and the entire body. The averages for the several organs are derived from observations on the different cases, so that the individuals from whom the liver was weighed are, for example, not always individuals from whom the brain also was weighed.

The figures for the males show the liver to be ultimately the heaviest organ, and the brain next in weight, although at birth the brain is by far the bulkiest of the entire group. It is found, however, that its relative growth is less, and very much less, than that of the other organs, and as one consequence the percentage decrease in the weight value of the brain is the greatest. The growth of the brain is most rapid during the first year. It will not escape notice that the maximum weight reached by the brain occurs in the table (in heavy type) for the males at fifteen years, and in that for the females, at ten years. This peculiarity will be discussed later on, and here it only need be remarked that there is no evidence for the view that in a given individual the brain at this period is heavier than it is during the next two decades. The most rapid weight increase of the other organs centres about the time of puberty, and tends to precede the maximum weight increase for the entire body; suggesting, therefore, that the muscular system is the last to respond. The relative growth of the left lung approaches most nearly that of the body as a whole, while the liver shows the greatest acceleration at puberty. Thus far, what has been said of the males applies to the females also.

There are a number of points, however, in which they differ. In the female at birth the brain, heart, right lung, and spleen are recorded as heavier than in the male, although the total body-weight is less. Both the relative and absolute increase in the male is greater for every organ except the spleen, while the percentage of the total body-weight as represented by those organs at maturity is less in all instances. The increase in the liver at puberty is most marked in the female. If we bring these facts together, it appears that the brain grows in a manner different from the viscera of the trunk, and that these latter more nearly follow the body-weight, though in the case of all, except the left lung, they fail to keep pace with it. Further, the acceleration of growth at puberty appears earlier in the female, and the percentage value of the viscera is greater. Owing to the smaller development of the muscular system in the female, this percentage relation is intelligible.

By way of criticism, it must be admitted that the original observations used in forming these tables were not homogeneous, and that for the most part the observations were made on persons belonging to the least favoured social classes. Further, that in view of the period covered, the number of observations for each organ is by no means large; yet, despite these evident deficiencies, there is reason to believe that the peculiarities here noted have a general significance.

Passing next to the examination of the increase in weight and size of the organs, as dependent on changes in their constituent elements, it will be necessary first to inquire whether, with the enlargement of the whole organ during this interval, the constituent elements increase in size only or in number also. For this inquiry the two methods of enlargement should be clearly understood. In describing abnormal growths, the pathologists employ the term hypertrophy, with the connotation that the enlargement thus designated is due solely to the increase in the size of the elements already formed. On the other hand, the term hyperplasia is used to indicate enlargement dependent on the formation of new cell elements. These terms are also convenient in distinguishing the two modes of normal growth. Taking the elementary tissues as previously enumerated, it is found that after birth the epithelial tissues in various localities are being continuously renewed though not increased, and that the connective tissues certainly exhibit hyperplasic increase throughout life. In the muscular system it is the unstriped muscles in which the tendency to hyperplasic growth is most marked and continuous; while this form of growth is not known to occur after birth among the nervous elements.

Should we be willing to take a suggestion from pathology, it would be a fair inference that those systems or organs in which hyperplasic growth was normally most marked, and continued through the longest time, would be the ones in which pathological forms of this process most frequently appear. The facts will, I believe, bear out such generalisation. The epithelia of the body surfaces are very variable, and adjust themselves to new conditions with ease. Within limits they regenerate, as in the case of small wounds, in the course of glandular activity, and the like. Of the examples of hyperplasic activity of connective tissue there would simply be no end, should an enumeration of instances be attempted. In many ways this least specialised of the tissues, possessing in a high degree the power of hyperplasic growth on slight stimulation, stands in the mammals not only as the prime element in all bodily repair, but as ever ready to assert its capability for increase even to the detriment of the organism as a whole. In many cases, no sooner does one of the specialised tissues grow weak and lose the control which each healthy cell-group exerts on its neighbours, than it becomes at once more or less overgrown by these vigorous but comparatively unspecialised connective tissue cells. The vascular and alimentary systems, together with the uterus during gestation, are the seats of the hyperplasic increase of unstriped muscular tissue. In the heart it is more difficult to determine the part which hyperplasia plays in the overgrowth, but so far as the skeletal muscles are concerned, hyperplasic growth as a normal occurrence is generally denied. Certain it is that ordinary exercise fails to increase the bulk of the muscles after a comparatively early period—namely, the twenty-fifth year, or the probable time of the cessation of growth, and that during the growing period the increase in weight is largely if not entirely the result of an increase in the size of the original fibres. If we are to judge from the activities of those elements, which by division give rise to the nerve cells, it is at a time some months before birth that the hyperplasic growth in the central nervous system of man terminates. But in attempting thus to sharply separate these two processes a false idea must not be conveyed. Strictly interpreted, hyperplasia implies the formation of new cells. Yet at birth, and for a long time after, many systems contain cell elements which are more or less immature, not forming a functional part of the tissue, and yet under some conditions capable of further development. The changes caused by the continued growth of these cells are not, strictly speaking, to be classed under either of the two heads just given, although for practical purposes organs enlarging in this way may be considered as undergoing hyperplasic increase.

It is evident from this brief survey how complex a process increase in weight may be: it may be due to a swelling of the cells already functional, the development of immature cells, or the production of new cells, and these processes may be combined in any proportion, while accompanying them in every organ is a variable increase in the connective tissue framework and the system of nutritive channels.

Associated with the study of these changes are two statistical difficulties of some importance. The observations based on the weight of portions of the body are made, and must necessarily be made, on the dead. So far as non-accidental causes of death are concerned, the anatomical peculiarities of the person dying are probably an important factor in causing death, for there can be no doubt that persons, the relation of whose various systems and organs are dissimilar, do also present different degrees of resistance to attacking disease or mechanical strains. An example of the latter is the observation by Boyd, that still-born children have heavier brains and larger heads than those born living, a result to be explained by the fact that under the mechanical conditions of birth a large head in the child is distinctly a disadvantageous variation. The dead, therefore, representing those individuals least well organised, form one series, and the living another. Where growth is very rapid, as in the lungs, slight variations may be obscured, and probably the curve from the dead does not materially mislead us as to the changes taking place in the living; but when we find that among the males in the table before us the average brain-weights are heavier at fifteen and twenty years than they are at twenty-five years, and among the females the same is true, with the addition that it holds at ten years also, the question at once arises whether there are here simply misleading figures, or whether the records are significant. Since other series of observations on the weight of the brain show the same pre-maxima, we can feel fairly certain that their occurrence is more than accidental. The social status of the persons on whom these observations were made was essentially similar at all periods, and the variation cannot therefore be explained by a change in that condition. Furthermore, this feature occurs in the statistics for both sexes, and, as we have noted, in the females at an earlier date than in the males. The suggestion is at least plausible that in these cases a premature growth of the brain has been one factor in causing death.

These facts can perhaps be taken as examples of the danger attending the deviation from the mean in the construction of a body, for in the case of the brain, as in that of any other organ or system, undue development weakens the entire individual, and so the facts derived from necropsies are to be applied to the history of living individuals only after proper correction. In connection with this last point the mortality records may be cited. Observations have shown that the percentage of deaths in the population at large varies with age, being high in the first years of childhood and in old age, and low in the middle years. These relations are expressed in Fig. 14.

The curve just given needs perhaps a word of explanation. It was developed by Hensen on the basis of Böckh's tables of mortality, which tables depended on Prussian statistics for the years 1865–6.[1] The tables are based on the assumption of a population of one million individuals, of the various ages from 0 to 100 years. The number of deaths occurring among those of a given age is represented as a percentage. The portion of the curve before 0 years represents the percentage of prematurely still-born children. It is easy to see that, while the rate of mortality at birth is very high indeed, it rapidly sinks until about the fifth year, when it has fallen to the neighbourhood of 1 per cent. From this age it rises slowly for the next sixty years, and finally very rapidly. An explanation of this is to be found by considering both how the body is built up and how it breaks down.

That the body is a house is a simile coming down to

[1] Hensen, Hermann's *Handbuch der Physiologie*, 1881.

us from the remotest antiquity, but we may enlarge upon it. Like a house, it is intended to resist destruction from the outside influences always acting upon it. If we interpret this mortality curve as an expression of the perfection of the bodily house, we find that during

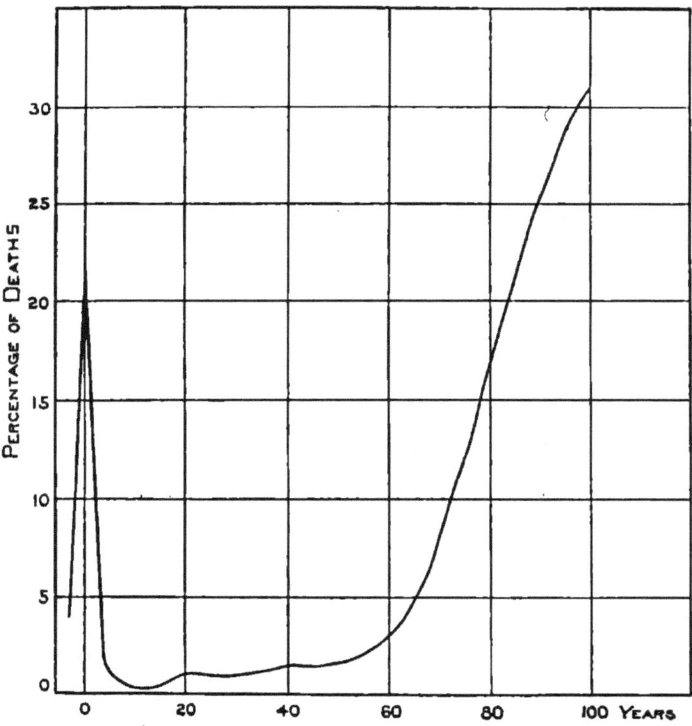

FIG. 14.—Mortality curve from Hensen, based on the figures of Böckh, 1876. The curve is intended to show the percentage of deaths occurring at each age. The calculations are based on the number of persons of that age present in a population of 1,000,000. (Hermann's *Handbuch der Physiologie.*)

the first five years of life, when, to extend the simile, the materials are in the process of being put together and adjusted to one another, the house offers but a small resistance to the forces which attack it. As we depart from the initial stages the more consolidated becomes the bodily structure, and the nearer the approach to a satisfactory *modus vivendi.* For a period of fifty-five years, from the age of five years onward, there is a strong and steady withstanding of these influences which had previously been so fatal. After this time ruin sets in, and the structural elements fall apart. From this there is no possible escape, for the good building material of sixty years before has in a large measure lost its strength and becomes rapidly less fit for the purposes of life.

Returning with these facts to the statistics for the weight of the entire body, derived by the general method, we see that account must have been taken of both those individuals who will successfully grow during the first five years and those who will not, whereas the later observations are made on those among whom the death-rate is much diminished. The character of the data, then, for the first five years of life is particularly heterogeneous.

At the latter end of the curves for increase both in weight and stature there is also a difficulty of importance. As regards weight, it is almost impossible in the years after twenty-five to determine when the proper growth ceases, and when increase in weight becomes due to the accumulation of body-fat, a habit in which different persons vary widely. Yet however strong may be the inclination to scorn fat as an ignoble tissue, and a disturbing factor in the proper study of growth, it is nevertheless not without its significance. The chubbiness of infancy, the lankness of youth, and the roundness of maturity are all normal phases of the body. From the physician's point of view this relation has been studied by Stephenson, whose curves for the absolute increase in weight were given in the last chapter; also it has been pointed out by Macauley[1] that in granting life insurance there is an increased risk if a person of given age and stature falls below a proportionate body-weight, for in such a case some morbid process, perhaps otherwise unrecognised, is at least to be suspected. In stature, too, it is very difficult to say when growth actually ceases, because possible correlations between stature and length of life come in to obscure the result. Baxter found in men an increase in stature up to the thirty-fifth year.[2]

It would be most natural to interpret growth processes as a genuine expression of bodily vigour, and it might be expected that during the period of growth the death-rate therefore would be small. We see from the mortality curve just examined that during the very active growing period of the first five years of life the mortality is comparatively high, but at that time, as we know, there are not only the active processes of enlargement, but also the necessary adaptation of the several systems to one another, and of the body as a whole to its environment. Should the term vigour be expanded to mean the capacity for the successful performance of all these processes, then the above interpretation would be admissible. Passing on to that period of growth which is associated with adolescence, it is shown by the observations of Key [3] that just after this has begun the percentage of illnesses among school-children decreases, and that the general health of scholars in Sweden is best during the middle of this period. With the cessation of the active growth the percentage of illness again rapidly rises.

The facts relating to stature are by no means as important for us as those bearing on body-weight,

[1] Macauley, *Quart. Public. of the Am. Statist. Assoc.*, 1893.
[2] Baxter, *Statist. Med. and Anthrop.*, Washington, 1875.
[3] Key, *Schulhygienische Untersuchungen*, Burgerstein, 1889.

but at the same time they possess much interest. Stature is mainly dependent on changes in the bony skeleton, or, to express it in a still more general way, in the connective tissue framework. In making measurements of it, then, we are following changes which take place in one of the great tissue systems of the body.

TABLE 8.—GIVING THE AVERAGE STATURE OF ALL CLASSES OF PEOPLE IN ENGLAND AT YEARLY INTERVALS. THE RECORD IS IN ENGLISH INCHES. THE MEASUREMENTS WERE MADE WITHOUT SHOES. (*From Roberts.*)

MALES. General population. All classes—town and country.			FEMALES. General population. All classes—town and country.	
Age last Birthday.	No. of Observations.	Average height in Inches.	Average height in Inches.	No. of Observations.
Birth	451	19·5	19·3	466
0—1	2	27·0	24·8	6
1	1	33·5	27·5	9
2	5	33·7	32·3	6
3	33	36·8	36·2	43
4	107	38·5	38·3	99
5	201	41·0	40·6	157
6	266	44·0	42·9	189
7	307	46·0	44·5	173
8	1524	47·0	46·6	432
9	2278	49·7	48·7	499
10	1551	51·8	51·0	480
11	1766	53·5	53·1	441
12	1981	55·0	**55·7**	225
13	2743	56·9	**57·8**	206
14	3428	59·3	**59·8**	240
15	3498	62·2	60·9	201
16	2780	64·3	61·7	136
17	2745	66·2	62·5	88
18	2305	67·0	62·4	62
19	1434	67·3	62·8	98
20	880	67·5	63·0	130
21	757	67·6	63·0	60
22	558	67·7	62·9	53
23	592	67·5	63·0	24
24	517	67·7	62·7	21

When stature is considered we find that the curve by which its increase is measured is in general similar to the curve for increase in weight, and that relations similar to those found for weight subsist between the curves for the two sexes. The most important difference between these two sets of measurements is that the periods of most rapid increase in stature precede those for weight. The increase in stature is due primarily to a lengthening of the skeleton in the line of its long axis; while the enlargement of the mass of muscles is especially connected with the increase in other diameters. At first glance it might appear that this growth in the long axis of the body occurs in the line in which gravity acts most strongly during the hours of activity, yet it is probable that the principal increase in stature occurs at night, during repose, when the long axis is horizontal. Moreover, the general impression to be gathered from the observations of other animals, and even of plants, is that where there exists a distinctly indicated long axis, this axis is the line of first growth, the growth at right angles to it coming later, and that gravity is not a factor of importance in this connection. The determination of the increase in stature before birth is complicated by the fact that it first involves the trunk alone, the limbs becoming important factors only later, hence there is no consensus as to the time when they should be included in the measurement. After birth the rate of increase in stature is comparatively slow. Its final cessation is indeterminate, although it certainly becomes very small after the twenty-fifth year. Fig. 15—formed in the same manner as Fig. 13—which shows rate of increase in weight, indicates that since it is a linear measurement the rate is always slower for stature, and that there is less difference between the first and later years of life.

All the curves which have been here presented are sinuous, thus indicating that periods of more rapid growth alternate with those in which it is slower. As has been pointed out, such a rhythm can be followed in the segmenting ovum. When the cells become more numerous the period of these alternations becomes longer. There is apparently an acceleration of growth processes at the sixth month of fœtal life, at birth, at the seventh year, and at adolescence, this being the last. During the growth period of the healthy child it becomes

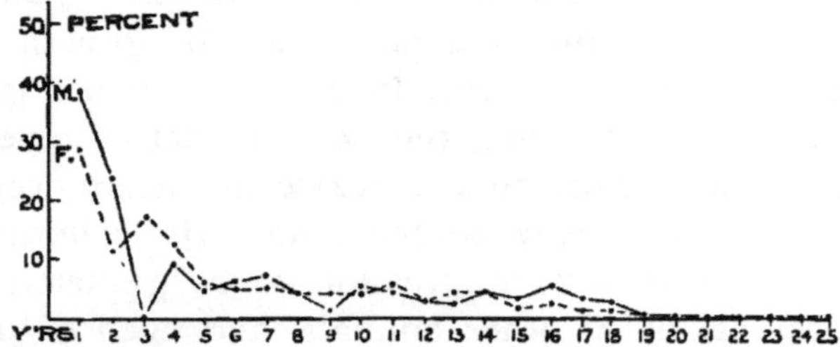

FIG. 15.—Curve to show the rate of increase in stature during the first twenty-five years of life. M. Males. F. Females. The percentages of increase are measured along the vertical axis, the ages along the horizontal one. (Based on Roberts, Table 8.)

of interest to inquire how the circling seasons and revolving day may influence the process. The observations of Malling-Hansen[1] on Danish children from nine to fifteen years of age show that by far the most rapid increase in stature was in the third of the year between the middle of April and the middle of August, while the third of the year between the middle of August and the middle of December was the one in which they gained nine-elevenths of their annual increase in weight. These observations have been made, however, upon

[1] R. Malling-Hansen, *Perioden im Gewicht der Kinder und in der Sonnenwärme*, Copenhagen, 1886.

school-children, to whom the summer season brought an increase in outdoor life and a respite from school work, as well as a change in meteorological conditions. Yet if season is here the important condition it follows that the children in the northern hemisphere have, like plants, a growing period the reverse of those in the southern ; but on this point there are as yet no observations. The various and careful observations of Camerer[1] have shown that a child of ten years is 700 grms. lighter and 2 cms. taller in the morning after a night's rest, and that during the day it is losing in stature and gaining in weight. Of course in this case not all the apparent gain in stature is lost during the following day, neither is it probable that the loss in weight at night reduces the total weight to the point at which it was the morning before, but it is extremely difficult, on account of the many sources of error, to directly determine these increments within such short intervals, and for these reasons we really know nothing of daily increase in weight or stature. With this review of general growth the matter must be left, in order that the more special questions relating to the nervous system may be taken up.

[1] Camerer, *Jahrb. f. Kinderh.*, Leipzig, 1893.

CHAPTER IV.

THE WEIGHT OF THE BRAIN AND SPINAL CORD.

Historical—Objects of examination—Interpretation—Constituent elements—Sources of error—Corrections—Reasons for not fusing results—Percentage of body-weight—Determination of brain weight in the living—Nomenclature—Spinal cord—Encephalon—Subdivisions—Grey and white matter—Chemical reaction—Percentage of water—Specific gravity—Weight of adult brain—Proportional development—Weight of the basal ganglia—Weight of spinal cord—Conclusions.

THE weight of the encephalon, or that portion of the nervous system contained within the cranium, has often been recorded, especially during the last century. The earlier observations were desultory and few in number, while the later ones are both more systematic, more accurate, and more numerous. Nevertheless very precise results are demanded, and until still greater care is taken it will not be possible by increasing merely the number of observations to pass much beyond the standpoint of to-day. In his *Elements of General Anthropology* Topinard [1] has given a list of the observations on the weight of the brain up to the year 1885. The entries in the table there printed show a total of 6,035 observations on the sane, and 4,147 on the insane. Since 1885 there have been recorded some 3,500 more, mainly on the insane, so that there now exists a grand

[1] Topinard, *Elements a'Anthropologie générale*, 1885.

total of over 13,000 observations. For the most part these have been made in England and in Germany, and and to a lesser extent in France, Italy, Russia, and Austria.

Investigators who have weighed the brain and its parts have had before them several objects. The facts were primarily of importance to the medical profession from the anatomical and anthropological point of view, while pathologists sought to associate different forms of mental disease with variations in the brain-weight and form. At the same time the search for a correlation between the size and form of the brain and the degree of the intelligence has interested all who have worked on this organ, and although it might be designated as the psychologist's standpoint, it has, from the very first, been in some measure before the minds of all. To a statement of the observations on the physical characters of the central nervous system we therefore turn.

Experiment and clinical observation have very clearly shown that disturbances of the encephalon are capable of causing disturbances of intelligence, and the study of the evidence justifies us in claiming the brain as the organ of the mind. Yet the brain is not entirely made up of nervous tissues, but contains other elements as well. It is surrounded by membranes, the pia and dura, supported inside and out by a framework of supporting tissues, and penetrated in all directions by nutrient channels, the blood-vessels and the lymphatics. This mass, consisting of nervous tissues, supporting tissues, and nutrient vessels in various degrees of distension, sometimes with the pia, sometimes without it, is the structure, the weight of which has been determined and recorded as that of the brain. Within the encephalon are cavities, the ventricles, and in these ventricles are to be found under different conditions varying quantities of fluid. When the brain is weighed with the ventricles unopened the fluid there present is included, and contributes to the final figure. In the examination of healthy brains the development of all these non-nervous factors, the fluids, membranes, and vessels, is either assumed as proportional in the several brains compared, or more rarely some effort is made to exclude them from the result by treating them separately. The dura weighed 40 and 42 grammes in the two cases, a woman and a man, in which E. Bischoff tested it.[1] Giacomini[2] found under ordinary circumstances that the weight of the pia and residual fluid was from 5 to 5·5 per cent. of the total weight of the hemispheres, these in turn being but 87·2 per cent. of the entire encephalon. If the vessels were congested the proportion might rise to 6·5 per cent. The most complete observations on the absolute weight of the pial membranes is furnished by Broca.[3] He found that the pia was heavier in the male than in the female, and that its weight increased with age. When the subjects obtained from a hospital for the insane, like the Bicêtre, were compared in this respect with those from the general hospitals, the pia was found to be heavier in the case of the insane. His table for the males is as follows:—

TABLE 9.—SHOWING THE WEIGHT OF THE PIA—MALES.
(*Broca.*)

20–30 years	... 45 grammes.
31–40 „	... 50 „
60– „	... 60 „

The variations ranged between 38 and 130 grammes,

[1] Bischoff, *Zeitschr. f. rationelle Med.*, 1863.
[2] Giacomini, *Guida allo studio dell' circonvoluzioni cerebrali dell' uomo*, Torino, 1884.
[3] Broca, quoted by Topinard, *Elements d'Anthropologie générale*, 1885.

both of which extreme observations were made on brains of small weight. The mean weight of the pia in the case of 133 females was found to be 48·7 grammes, and in the case of 273 males 55·8 grammes. Th. v. Bischoff has determined the weight of the pia as from 25 to 40 grammes.[1] In the insane Morselli [2] finds the weight of the pia and fluids to be almost double that found in normal persons.

The cast of the ventricles as made by Welcker displaces 26 cu. cm. of water, so that the fluid filling such a cavity would weigh a trifle over 26 grammes. From these results some notion of the part played by the ventricular fluid and the pia is to be obtained. There are numerous other records relating to the dura, but they differ so widely that evidently they should not be compared with one another. Variations in the proportion of these non-nervous constituents can produce noticeable discrepancies when single brains are compared, but as soon as average weights based on series of twenty or more specimens are taken the disturbance caused by them ceases to be important.

At one time it was my purpose to combine all existing data, and from the broader basis thus established to test the validity of conclusions current in this field. It soon became plain, however, that for a number of reasons the observations made by different observers using dissimilar methods and working with various races, were not in themselves homogeneous enough to bear fusing without very important corrections. It has therefore been deemed best to discuss the question before us simply by the aid of a few of the best series of observations.

In Table 5, already given, the brain is recorded as

[1] Bischoff, *Das Hirngewicht des Menschen*, 1880.
[2] Morselli, *Rev. de l'Anthropologie*, t. 1, 1890.

weighing from 2·16 per cent. (male) to 2·23 per cent. (female) of the entire adult body. So long as it is intended to use these figures merely to indicate that the proportional weight of the brain is small, no harm is done, but, given the body-weight, it is quite unjustifiable to attempt to deduce the weight of an individual brain by the aid of it. A moment's thought will show that the chief variable in the original determination of the percentages was the body-weight. The chief variable in this proposed inference is the body-weight, and the variation in that figure being wide, the final result must be correspondingly unreliable.

Although the problem has been repeatedly attacked, it must be admitted that there is no method of determining with satisfactory accuracy the weight of the brain in the living person, so that for the present at least, the facts obtained by autopsies supplemented by the determination of the capacities of skulls, are alone useful. In studying the questions connected with the weight of the brain it will be necessary to employ descriptive terms which are exact, in order to understand just what has been weighed, and a word on these terms is therefore in place.

The nervous system is divided into a central and a peripheral portion. Since there is no natural line of division between these two, an arbitrary separation is made at the point where the nerves leave the cavity of the skull and vertebral column. Of the weight and volume of the peripheral nervous system which ramifies through all parts of the body, records are wanting, because of the difficulty in separating it from the surrounding parts. But from the cranial cavity and the spinal canal it is comparatively easy to remove the enclosed masses, and thus the brain and spinal cord have been weighed in various ways. To dispose first of the portion about which our information is most meagre, it may be said that we know very little concerning the physical properties of the spinal cord, a lack for which the comparative difficulty of obtaining the cord for examination is to a large extent responsible. It is always weighed after the removal of the dura, but with the pia adherent. It should be added, however, that the weight of the cord also varies with the manner in which the nerves are separated from it. Since the nerves arising along the cord may pass for several inches within the canal before going through the vertebral foramina, it becomes possible to separate them either at the point where they arise from the cord or at the point where they enter the foramina. Having a considerable mass they add materially to the weight of the cord, which therefore varies accordingly as it is weighed with or without them. Occasionally the cord is farther divided into the portions designated as the cervical, thoracic, and lumbar, and these weighed separately.

At the base of the skull the portion of the central nervous system within the cranium contracts to form the bulb, which is directly continuous with the spinal cord, and one of the first difficulties is to determine the proper point of separation between these two. The mere acceptance of the mass within the cranial cavity as the encephalon does not satisfy the conditions, because both the brain and the spinal cord are capable of some longitudinal movement, and as a result the amount which may be included within the cranial cavity is slightly variable. It has consequently become the custom to select as a fixed point the caudal end of that elevation on the ventral surface of the bulb, known as the decussation of the pyramids, and to make the section at this level. The mass above this section is the encephalon. This also establishes the caudal boundary of the bulb. Its cephalic boundary is taken at its junction with those great bundles of transverse fibres which form the pons. This region of the pons has in turn its anterior boundary indicated by a plane passing through its cephalic edge and separating it from the corpora quadrigemina. Between this boundary and a plane passing through the cephalic edge of the corpora quadrigemina is cut off the fraction of the encephalon known as the mid brain, but it is only rarely that this portion is weighed alone, being usually left in connection with the hemispheres. The hind brain, or cerebellum, is connected on each side with the bulb, pons, and the region of the corpora quadrigemina. When all these connections are cut, the cerebellum is separated, and may be weighed either *in toto* or after further division. The term cerebrum is variously applied to that portion of the encephalon which lies in front of either the cephalic edge of the pons or of the corpora quadragemina, authors having adopted different usages. It will thus be seen that the weight and volume of the cerebrum must be somewhat greater in cases where the quadrigemina are still in connection with it. In the case of the tables from Boyd, which I shall present later, the term cerebrum means the brain mass cephalad of the quadrigemina. When the cerebrum is divided in the median plane, it is separated into two symmetrical portions, the hemispheres. To follow further the subdivisions which are sometimes made, it will suffice to consider one of the hemispheres alone. If the island of Reil, a sunken portion of the hemisphere, be exposed by turning back the opercula which cover it, and which are formed by the edges of the parietal, frontal, and temporal lobes coming together along the line of the Sylvian fissure, the part thus uncovered is seen to be bounded by a continuous sulcus. If a scalpel be carried through this sulcus cutting deeply, the hemisphere is separated into two portions, a smaller basal portion composed of the basal ganglia (optic thalamus and corpora striata, together with the portion of cortical substance covering the striata and forming the island) and the more massive enveloping portion, the mantle.

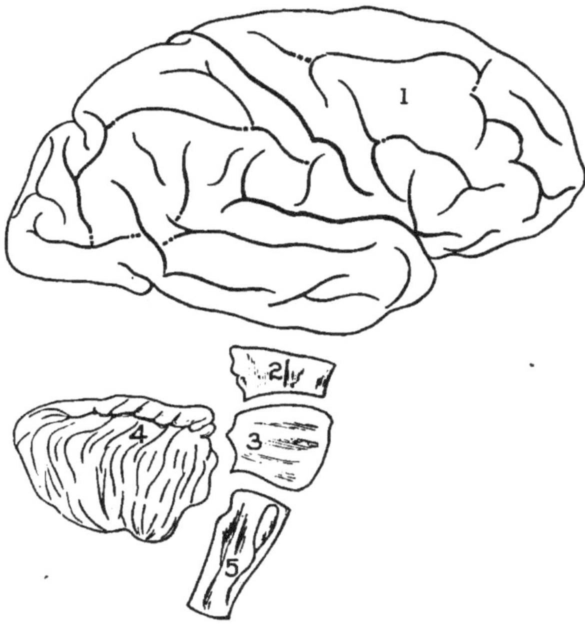

FIG. 16.—Showing the principal divisions of the encephalon made for the study of its weight :—1. Hemisphere, seen from the side, fissuration according to Eberstaller; 2. Mid-brain, region of the corpora quadrigemina; 3. Pons; 4. Cerebellum, or hind-brain; 5. Bulb, or after-brain. Parts 2, 3, and 5, taken together, form what is designated the "stem" in the tables of Boyd. (Modified from Quain's *Anatomy*.)

The mantle consists of a central mass of white matter completely covered over with grey cortex. Sometimes the mantle is still further divided into lobes, but the value of these subdivisions and the extent to which they have been employed can be best discussed when the occasion arises. Fig. 16 is intended to show the principal subdivisions which have just been enumerated.

The nerve tissue proper is composed of two portions distinguishable by their colour. Certain parts of the central and the peripheral system appear grey in colour, while the remaining portions are white. The white matter is composed chiefly of medullated nerve fibres, and owes its colour to their white medullary sheaths. The grey matter, on the other hand, contains the bodies of nerve cells, but fibres arising from them are present in large numbers, thus making the grey substance less homogeneous. To the relative amount of white and grey we shall later return. In their normal and resting condition both white and grey matter are alkaline in reaction, but after death they tend to become neutral or even acid. Like all the soft tissues the nervous system contains a large proportion of water. The tables which follow give the percentage of water found in the grey and white matter of the brain and in the spinal cord.[1]

TABLE 10.—SHOWING THE PERCENTAGE OF WATER IN THE NERVOUS SYSTEM. (*From Halliburton.*)

LOCALITY.	MONKEY.	DOG.	CAT.	MAN.
Grey Matter of Cerebrum	81·8	82·1	82·3	85·5
White Matter of Cerebrum	70·0	70·3	69·2	69·6
Cerebellum, Grey and White together	—	—	78·7	80·4
Spinal Cord as a whole, Grey and White together	70·2	70·0	70·8	74·8
Cervical Cord	70·1	71·6	71·7	74·7
Thoracic Cord	66·5	68·3	69·0	74·1
Lumbar Cord	74·1	70·1	71·8	75·7
Sciatic Nerves	—	—	61·3	—

This table shows a striking similarity in the percentage of water found in animals widely separated from one

[1] Halliburton, *Journal of Physiology*, 1894.

another in the zoological scale. As has been previously explained, the white matter is more homogeneous than the grey, and the percentage of water in the white matter therefore shows the least variation. The order in which the animals stand in this table was determined by putting first the one with the smallest percentage of water in the grey matter of the cerebrum and letting the others follow in regular succession. Passing from the monkey to the dog, cat, and man, the percentage of water increases in the other portions of the nervous system much as it does in the grey matter of the cerebrum, since the proportion of cell-bodies increases.

In looking at the subdivisions of the spinal cord as made for this examination, it is to be remembered that proportionately the cervical and lumbar regions contain more grey matter than the thoracic region does, and as a result both those regions show the greater percentage of water. Since in the lumbar enlargement the proportion of white matter is less than in the cervical, we should expect, as the records show, that it would contain the greatest proportion of water. Using a similar method to that of Halliburton, De Regibus obtained in Italian brains the following figures, which apply, however, to the cortex and white matter of the cerebrum only.[1]

TABLE 11.—SHOWING OBSERVATIONS BY DE REGIBUS (GIACOMINI) ON ITALIAN BRAINS. THE FIGURES INDICATE THE PERCENTAGE OF WATER IN THE GREY AND WHITE MATTER OF THE MANTLE OF THE CEREBRUM.

PERCENTAGE OF WATER IN	CASE 4.	CASE 5.	CASE 6.	CASE 7.	AVERAGE
Cerebral grey matter	86·3	85·8	85·8	85·9	86
Cerebral white matter... ...	70·1	70·4	70·4	70·3	70·3

[1] *Vide* Giacomini, *Guida*, &c., Torino, 1884.

These in turn agree very closely with the observations made by Thudichum on English brains,[1] and which give:—

Percentage of water lost at 95° C., from cortical grey matter, 85·27 per cent; percentage of water lost at 95° C., from white matter of hemisphere, 70·23 per cent.

Taken all together, these results indicate in a satisfactory way the constancy of the proportion of water in the parts named. The variations which occur are readily explained by slight differences in the localities from which the samples tested were obtained and by errors of experiment.

Directly connected with the percentage of water is the specific weight of these substances. Obersteiner[2] has made some very careful observations on the specific gravity of the cerebral cortex from the different portions of the hemispheres, and finds a striking and significant increase in this figure on passing from the frontal to the occipital lobes.

TABLE 12.—SHOWING THE SPECIFIC GRAVITY OF THE ENCEPHALON AT DIFFERENT POINTS. (*Obersteiner.*)

F.—Frontal lobe; P.—Parietal lobe; O.—Occipital lobe; T.—Temporal lobe; C. str.—Corpus striatum; Th. O.—Thalamus Opticus; C. dent.—Corpus dentatum.

CEREBRAL CORTEX. *Right Hemisphere.*			*Inter-Brain.*			CEREBRAL CORTEX. *Left Hemisphere.*		
F.	...	1·0308	White Matter		1·0412	1·0308	...	F.
P.	...	1·0325	C. str.	...	1·0378	1·0325	...	P.
O.	...	1·0362	Th. O.	...	1·0402	1·0360	...	O.
T.	...	1·0326				1·0330	...	T.
			Cerebellum.					
			Cortex	...	1·0376			
			White Matter		1·0412			
			C. dent.	...	1·0400			
			Pons	...	1·0413			
			Bulb	...	1·0371			

[1] Thudichum, Tuke's *Dictionary of Psychological Medicine*, 1892.
[2] Obersteiner, *Centralblatt f. Nervenheilkunde*, 1894.

What is still more important is the fact that the specific gravity of the cortex itself increases from the surface inwards towards the white matter, and if it be divided into three layers the following is found :—

Outermost layer	...	...	Specific gravity,	1·028
Middle layer	...	...	„	1·034
Innermost layer	...	...	„	1·036

It will be seen from these figures that before the various records of the proportion of water in the grey matter of the cortex can be fairly compared more detail is required concerning the methods of obtaining the samples examined. From a selection of the best facts available, I have calculated that the average specific gravity of the entire encephalon should be for the adult male, 1·0363, and for the adult female, 1.036. In a case, the sex of which is not given, Thudichum by direct observation obtained for the entire encephalon a specific gravity of 1·0373.

It will be found, I believe, that these determinations are very close to the truth, for the case of individuals in the normal condition and in the prime of life, and that most of the wide variations which have been reported in this connection have been due rather to the method of experiment than to differences inherent in the specimens. Turning now to the weight of the brain itself, we find a large number of average weights on record, but such figures are of little interest since the "average man," to whom they are assumed to belong, is a myth. It is far more advantageous to subdivide the records into smaller groups, according to the conditions which are found to be important, and to make the averages for these special groups, since observers are now agreed that age, sex, stature, body-weight, and race are all modifying circumstances. The most complete single series of observations is that which was made in

England by Dr. Boyd.[1] In a recently published paper by the late Dr. John Marshall[2] particularly good use has been made of this material, and the presentation of the facts there will best serve our purpose. Since Boyd's series was made on English subjects, it is more comparable than any other with the tables for the growth of the body which have been already given. Dr. Boyd examined the brains of 2,086 individuals, presumptively sane, at the Marylebone Workhouse in London. His method of observation is not completely stated, but was apparently as follows :—The brain was exposed, the pia being left in place. By horizontal sections the hemispheres were sliced away down to the tentorium. The remaining portions of the hemispheres were then separated from the quadrigemina. The cerebellum was next separated from the stem, represented by the quadrigemina, pons, and the bulb. Each hemisphere, the cerebellum and the stem, were weighed separately. It will be seen that this method permitted very complete drainage of the fluids found in the brain after death, and this must be taken into account when Boyd's figures are compared with the tables of some other observers who omitted this precaution. Dr. Marshall compiled from the records of Boyd the observations which were important for his purpose, and on these we must depend, since the separate observations by Boyd were never published and are now lost. Between the ages of 20–90 years, Marshall finds 698 male cases and 552 female. In making the table he divides these cases into three (horizontal) groups according to age, putting together those from 20–40, those from 41–70, and those from 71–90 years. The individuals who fall within each group of years are

[1] Boyd, *Phil. Trans.*, 1861.
[2] Marshall, *Journal of Anatomy and Physiology*, 1892.

again divided into three series according to their stature, thus giving us the figures for the tall, medium, and the short persons. Under each one of the classes thus marked out—that is, among people between given ages and of a given stature—are recorded the weights of the subdivisions of the encephalon as above named. There is thus obtained an unusually complete presentation of the weight as affected by the most important conditions that modify it. Dr. Marshall's tables were expressed in English weights and measures, but these I have taken the liberty of translating into the metric system, since by far the greatest number of observations on the brain are thus recorded.

TABLE 13.—SHOWING THE WEIGHT OF THE ENCEPHALON AND ITS SUBDIVISIONS IN SANE PERSONS, THE RECORDS BEING ARRANGED ACCORDING TO SEX, AGE, STATURE. (*From Marshall's tables based on Boyd's records.*)

a. indicates that a record considered according to age is too large. s. indicates that a record considered according to stature is too large.

SANE.									
MALES.					FEMALES.				
Ages.	Encephalon.	Cerebrum.	Cerebellum.	Stem.	Stem.	Cerebellum.	Cerebrum.	Encephalon.	Ages.
Stature 175 cm. *and upwards.*					*Stature* 163 cm. *and upwards.*				
20-40	1409	1232	149	28	23	134	1108	1265	20-40
41-70	1363	1192	144	27	23	131	1055	1209	41-70
71-90	1330	1167	137	26	24 a	130	1012	1166	71-90
Stature 172-167 cm.					*Stature* 160-155 cm.				
20-40	1360	1188	144	28	26 s	137 s	1055	1218	20-40
41-70	1335	1164	144	27	26 s	131	1055	1212 s	41-70
71-90	1305	1135	142 s	28 a s	24	128	969 s	1121	71-90
Stature 164 cm. *and under.*					*Stature* 152 cm. *and under.*				
20-40	1331	1168	138	25	24 s	130	1045	1199	20-40
41-70	1297	1123	139 a	25	25 a s	129	1051 a	1205 a	41-70
71-90	1251	1095	131	25	25 a s	123	974	1122	71-90

On examining this table we find the following facts for consideration. In the case of the males the weight of the encephalon regularly decreases with each period of years. It decreases also when cases belonging to the same groups, according to age, are compared according to their stature, the tallest individuals having the heaviest encephalon and the shortest the lightest. Examining next the different subdivisions of the encephalon, it appears that except in a few instances the same general statements hold good. Where an entry does not have the value anticipated when the brains are compared according to stature, the fact is indicated by the insertion of a small s, and where it fails to conform when the comparison is made according to age it is indicated by a small a. Thus to be explicit, in the case of males of intermediate stature the figures for those between 71–90 years do not conform in the case of the cerebellum and the stem, whereas among the shortest individuals those between 41–70 years fail to conform in the case of the cerebellum alone. In the case of the figures for the females it is found that the same general statements hold good, save that the number of exceptions is larger. These latter are indicated in the same way, and it is hardly necessary to enumerate them in detail. As has been stated the observations on the females are based on a smaller number of cases, and it is especially in the groups where the number of individuals is probably least that the deviations of the general relations occur. We may conclude from these tables that after maturity the encephalon in the female is smaller than in the male, and that all its parts are smaller. When the observations are grouped according to age, they show that with increasing age there is a decrease in weight of the encephalon and in all its parts; that for those of the same age and sex decrease in stature is accompanied by a decrease in the weight of the encephalon and in all its subdivisions, and that in these respects the two sexes are similar. I have also recast the figures of Marshall in such a way that in the above table the weight of the encephalon is in each instance made the standard, being taken as equal to 100, and the percentage weights of the constituent subdivisions are then calculated. There is a remarkable constancy in the percentage values of the subdivisions of the encephalon of all ages, all statures, and both sexes. The value of the cerebrum in the female is in most of the averages less than 0·4 per cent. below that in the male, and this of course raises the percentage values of the other subdivisions, but whether any real importance is to be attached to this difference is debatable, although, so far as it goes, it is, in my opinion, significant. Advanced age in the female is also associated with a decrease in the proportional weight of the cerebrum. In Table 14 the averages have been made according to age, stature being neglected, and in Table 15, according to stature, age being neglected. The constancy in the proportions of the subdivisions of the encephalon is the most important fact thus demonstrated.

TABLE 14.—SHOWING THE PERCENTAGE OF WEIGHT OF THE SUBDIVISIONS OF THE ENCEPHALON, THE RECORDS BEING GROUPED ACCORDING TO AGE. BASED ON TABLE 13.

MALES					FEMALES.				
Age.	Encephalon.	Cerebrum.	Cerebellum.	Stem.	Stem.	Cerebellum.	Cerebrum.	Encephalon.	Age.
20–40	100	87·52	10·49	1·91	1·96	10·9	87·13	100	20–40
41–70	100	87·00	10·6	1·94	2·02	10·8	87·14	100	41–70
71–90	100	87·33	10·6	1·98	2·11	11·16	86·4	100	71–90

TABLE 15.—SHOWING THE PERCENTAGE OF WEIGHT OF THE SUBDIVISIONS OF THE ENCEPHALON, THE RECORDS BEING GROUPED ACCORDING TO STATURE. BASED ON TABLE 13.

MALES.					FEMALES.				
Stature.	Encephalon.	Cerebrum.	Cerebellum.	Stem.	Stem.	Cerebellum.	Cerebrum.	Encephalon.	Stature.
175 cm. and upwards.	100	87·5	10·5	1·90	1·91	10·86	86·93	100	163 cm. and upwards.
172–167 cm.	100	87·2	10·65	2·08	2·10	11·16	86·68	100	160–155 cm.
164 cm. and under.	100	87·17	10·6	1·86	2·09	10·83	87·06	100	152 cm. and under.

The constancy of the proportional development has been heretofore insufficiently emphasised, but it is a matter that must be recalled when the significance of variations in the size of the encephalon is later discussed.

Franceschi[1] is the only investigator who has taken the trouble to study the combined weight of the corpora-striata and the thalami-optici. His method of preparing this subdivision was to excise the basal ganglia with the cortex of the insula attached, and then with a scalpel dissect away both the cortex and the white matter of the insula until the grey substance of the ganglia could be seen clearly. The figures here given are condensed from his Table 23. They show a constant difference between the sexes, similarity of the right and left sides, but no decrease with advanced age.

[1] Franceschi, *Sul peso dell' Encephalo*, &c., Bull. d. Sc. Med. di Bologna, 1888.

TABLE 16.—GIVING THE WEIGHT OF THE BASAL GANGLIA IN THE TWO SEXES AT DIFFERENT AGES. WEIGHT IN GRAMMES. (*Franceschi.*)

MALES (Basal ganglia).				FEMALES (Basal ganglia).			
Age.	No. of Obs.	Mean Weight.		Age.	No. of Obs.	Mean Weigh	
		Right.	Left.			Right.	Left.
Before 5 years.	2	25	24	—	—	—	—
21–40	16	41·2	40·8	21–40	20	36·0	36·0
41–70	38	41·6	42·3	41–70	45	37·7	38·0
71–87	22	42·4	42·4	71–87	21	37·7	41·0

The most comprehensive investigations on the weight of the spinal cord are those which have been recently published by Dr. Mies.[1] The human cord without its nerve roots was found in thirteen grown persons to weigh between 24 and 33·3 grammes, the intermediate group of four individuals showing a cord weight of from 25 to 27 grammes. This corresponds well with the figures given by E. Bischoff and others. So far as the data go, which is not far, to be sure, there is no distinct variation in the weight of the cord due to sex. The cord at the same time does not appear to be so closely correlated with body-weight as with stature, and it will be readily seen that an increase in the length of the spinal canal would naturally be accompanied by an increase in the length of the cord, and this in turn by an increase in its weight.

The general results which have been obtained from the examination of these figures gathered by Dr. Boyd are sufficiently well supported by other series of observations to justify their acceptance without further evidence in their behalf. It will be noticed that this discussion

[1] Mies, *Neurolog. Centralblatt*, 1893.

was opened by examining the weight of the brain after the twentieth year, when the most active growth changes have ceased. This was done since it seemed desirable to make the conditions found in the adult the point of departure, but the changes accompanying the increase of weight and volume during the first twenty years of life are still to be discussed.

CHAPTER V.

INCREASE OF THE BRAIN IN WEIGHT AND VARIATIONS OF THE CRANIUM IN CAPACITY.

Table and chart of increase in brain-weight with age—Weight at birth—Weight in the still-born—Early growth—Proportional growth—Spinal cord—Changes in composition—Cessation of growth—Venn's observations—Length and breadth of head—Brain-weight in different races—Cranial capacity—Determination of exact weight.

THE increase in the weight of the brain during the earlier years of life is an important change, for which the most complete observations are those furnished by Vierordt in the accompanying table.[1] In this table the two sexes are considered separately, and the growth of the brain indicated by recording its average weight in persons dying at the respective ages; though in some instances it will be noticed that the number of cases for a given year is much smaller than is desirable for the best results. The curves into which these figures have been thrown for graphic representation show decided irregularities, probably due to this same fact: nevertheless it is also true that the general character and relations of the two curves are not disturbed by these defects, and that certain broad conclusions may be safely drawn from them.

[1] Vierordt, *Arch. f. Anat. und Physiol.*, 1890.

TABLE 17.—TO SHOW THE INCREASE IN BRAIN-WEIGHT WITH AGE. ENCEPHALON WEIGHED ENTIRE WITH PIA. (*Compiled by Vierordt.*)

MALES.			FEMALES.	
Age.	No. of Cases.	Brain.	Brain.	No. of Cases.
0 Months	36	381	384	38
1 Year	17	945	872	11
2	27	1025	961	28
3	19	1108	1040	23
4	19	1330	1139	13
5	16	1263	1221	19
6	10	1359	1265	10
7	14	1348	1296	8
8	4	1377	1150	9
9	3	1425	1243	1
10	8	1408	1284	4
11	7	1360	1238	1
12	5	1416	1245	2
13	8	**1487**	1256	3
14	12	1289	**1345**	5
15	3	**1490**	1238	8
16	7	1435	1273	15
17	15	1409	1237	18
18	18	1421	1325	21
19	21	1397	1234	15
20	14	1445	1228	33
21	29	1412	1320	31
22	26	1348	1283	16
23	22	1397	1278	26
24	30	1424	1249	33
25	25	1431	1224	33
Total No. of Cases, 415			Total No. of Cases, 424	

At birth the weight of the encephalon is nearly alike in the sexes, and in both growth during the first year, and indeed during the first four years, is rapid. By the seventh year the encephalon has reached approximately its full weight, the subsequent increase being comparatively small. There is no other peculiarity in the growth process of either sex, unless later observations should show that the approximation of the curves at fourteen years was really significant. Again, the pre-maximal rise in weight is to be noted, but its meaning

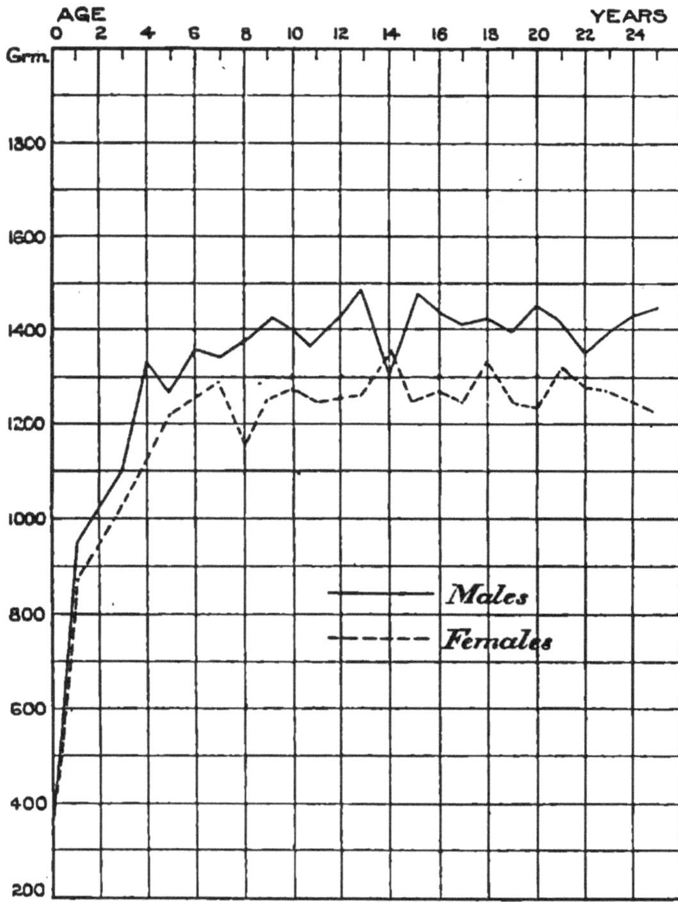

FIG. 17.—Curves showing the variations in brain-weight during the first twenty-five years of life. Based on Table 17.

has already been discussed. Should this curve be extended to ninety years, there would be found nearly the same weight of the brain persisting up to the onset of old age (about fifty years), when there appears a loss in weight, which becomes rapidly more evident, so that the smaller brain-weight of the aged must represent a percentage of loss in some instances quite large. For comparison it would be of interest to know how the cranial capacity of the extremely aged compared with that of those in the prime of life, and also whether there was possibly any correlation between longevity and the initial weight of the brain, for it is conceivable that a heavy brain, though developed late in life, may in itself be unfavourable to length of years. From the curves it is clear that the brain-weight for the females runs almost from the start below that for the males, that the difference increases during the period of most active growth, and at the end of that time reaches nearly its maximum. This difference is maintained throughout life.

Concerning the weight of the encephalon in males and females at birth, a word more may be added. Most of the older observers found at this period the greater weight in the males. These figures of Vierordt show a slight excess in favour of the females, and this same occurs in other recent series. Of course the condition of greatest importance in determining the weight of the brain at birth is the size of the child, which can vary, as we know, within wide limits. To the larger and heavier children belong, on the average, the heavier brains, and it will not be possible to clarify our ideas concerning the weight of the brain at birth and the influence of sex upon this weight until it is possible to compare more accurately male and female children, and to be sure that still-born children have not been included in the record. The matter is more complicated, and consequently more interesting, than at first sight appears. The records made by Boyd show the following relations :—

TABLE 18.—THE AVERAGE WEIGHT OF THE ENCEPHALON IN GRAMMES IN CHILDREN STILL-BORN AND IN THOSE BORN LIVING. (*Condensed from Boyd.*)

		MEAN BRAIN-WEIGHT.			
Still-born	Males	397	355	Males	Born living
Still-born	Females	352	287	Females	Born living

It is a suggestive fact that the greater part of the growth of the brain takes place before any of the formal educational processes have begun, for the mild schooling that occurs before the age of seven or eight years can hardly have much influence. It may be, perhaps, that such a slight increase in weight as occurs during the twenty years that follow is so distributed as to be much more effective than we at present imagine, but to this suggestion I shall return in a later chapter.

Exact measurements support common observation in the conclusion that the proportional development of the body is different at maturity from that at birth, and in the same way the proportional development of the encephalon changes during the period of growth. The most recent figures are given in Table 19. These observations were made by Danielbekof on two hundred Russian children, whose average age was one month.[1] The second table is based on the observations of Boyd.[2]

[1] Danielbekof, *Materialien zur Frage über das Volumen des Gehirns und der Medulla oblongata bei Kindern beiderlei Geschlechtes*, Inaug. Diss., St. Petersburg, 1885 (Russian).

[2] Boyd, *Phil. Trans.*, 1861.

TABLE 19.—SHOWING THE WEIGHT OF THE ENCEPHALON AND ITS PARTS, AND OF THE SPINAL CORD, IN 200 CHILDREN HAVING AN AVERAGE AGE OF ONE MONTH. METHOD OF WEIGHING NOT KNOWN. (*Danielbekof.*)

	MALES.		FEMALES.	
	Weight—grms.	Per-centage.	Per-centage.	Weight—grms.
Encephalon	415·3	100	100	399·2
Two Hemispheres ...	381·5	91·9	91·7	365·7
Cerebellum	28·1	6·8	7·0	28
Pons and Bulb	5·6	1·3	1·3	5·5
Spinal Cord	3·9			3·8

TABLE 20.—SHOWING THE WEIGHT OF THE ENCEPHALON AND ITS PARTS AT DIFFERENT AGES. (*Boyd.*)

The "stem" includes the bulb, pons, and quadrigemina, and is therefore not directly comparable with the "pons and bulb" of Table 19. The relative values of the subdivisions are thereby slightly altered. (*Topinard.*)

No. of Cases.	Age.	Cerebrum.	Cerebellum.	Stem.
		MALES.		
45	New-born	92·4	5·8	1·60
22	7 to 14 years	87·8	10·3	1·61
99	30 " 40 "	87·3	10·6	1·98
95	70 " 80 "	87·0	10·7	2·09
		FEMALES.		
45	New-born	92·1	6·2	1·50
18	7 to 14 years	87·9	10·5	1·50
80	30 " 40 "	87·0	10·8	2·01
128	70 " 80 "	86·9	10·9	2·15

The fact that in Boyd's Table (20) the stem includes more than the pons and bulb, prevents a direct comparison of the entries for the weight at birth with those in Table 19. But both tables emphasise the great development of the hemispheres at this time.

Boyd's table further brings out the interesting fact that there is comparatively small change in the percentage relationships after the period of rapid growth is completed—that is, after the twentieth year. The failure of these proportional figures after the twentieth year to exactly agree with those already given in Table 14, is probably due to the fact that the time intervals have been differently chosen. Indeed, a comparison of the two tables may serve to show, by reason of the small amount of differences, how little influence upon the proportional development is exerted by those conditions which have so much effect on absolute weight. It is further to be noted that in these latter tables the proportional weight of the hemispheres is as often superior in the female as it is inferior, hence we must be very cautious about inferring the influence of sex in this relation.

Without giving special figures, Mies states that the spinal cord in man increases in weight rapidly during the first years of childhood, and more slowly later. He further adds that it still continues to increase after the brain has ceased to grow, and that its senile atrophy begins at a later age than in the case of the brain. His published observations show that in the new-born child the weight of the cord ranged from 2 to 6 grammes, with a mean weight of 3·42 grammes. Table 19 shows that Danielbekof found the weight of the spinal cord in male children averaging in age one month 3·9 grammes, and in the female 3·8 grammes. Comparing these with Mies' figures for the weight in the adult (26 grammes) it appears that between birth and maturity the weight of the cord has increased over 7½ fold, as against an increase in the weight of the brain of less than one-half that amount. The proportional relations of the encephalon to the cord at different ages are very interesting. In the case of a three months' fœtus the brain weighed 18 times as much as the cord; in the fœtus of five months it was 101 times heavier; and in the case of 21 children born at full time (11 females and 10 males) it was almost 115 times heavier. On the other hand, in 10 men at maturity the brain weighed 51·3 times the weight of the cord, and in 4 females the proportional weight was 49·47. There are, however, no published figures concerning the growth of the cord during the first twenty-five years, so that this process cannot be compared in detail with the growth changes taking place in the brain. But from the foregoing it is to be seen that in proportion to that of the brain the increase in the weight of the cord in early fœtal life is rapid, at the time of birth slow, and then more rapid up to maturity.

With the increase of the central nervous system, occur changes in its composition concerning which a few observations exist. In its first form the entire central system is grey, and only later do portions of it become white by the development of myeline in the sheaths of the nerve fibres. The myeline contains fatty substances and a smaller proportion of water than the axial part of the fibre which it ensheaths. The following table indicates the proportion of water and myeline (extracted by ether) found in the white matter (fibres of the callosum) at birth and at maturity.[1]

[1] Schlossberger, *Ann. der Chemie u. Pharmac.*, 1853.

TABLE 21.—COMPARING THE PROPORTION OF WATER AND OF SUBSTANCES EXTRACTABLE BY ETHER IN THE FIBRES OF THE CALLOSUM OF THE NEW-BORN AND THE ADULT. (*Condensed from the Observations of Schlossberger.*)

PERCENTAGE OF WATER AT BIRTH.		IN THE ADULT.
89·48	....	70·60
89·60	....	70·60
89·79	....	70·68

PERCENTAGE OF ETHER EXTRACT AT BIRTH.		IN THE ADULT.
3·85	....	15·41
3·78	....	15·03
3·78	....	15·32

Observations made on the growth of any part made by the "general" method are of little use in determining the age at which growth in the individual ceases, because, so long as any of the individuals grow the curve will rise, and there is no way to determine whether the rise is due to a slight growth in many, or a greater growth in a few. Further than this it is not probable that the cessation occurs at the same age in all classes of the community, and we might fairly expect that it would be continued for the longest time in those most favoured. To this conclusion the observations by Venn in England directly point.[1] Suffice it to say that Venn measured the heads of students at Cambridge throughout the entire student period, and found that the head increased in height, breadth, and length during this time. From this it is inferred that the brain also increased, and it is highly probable that such is the case. The observations of West, made in the Worcester (Mass.) Schools,[2] show a steady growth in the length and breadth of the head between the ages of five and twenty-one years. They also show between the ages of eleven and thirteen years an approximation of the curves

[1] Venn, *Nature*, 1890.
[2] West, *Archiv. f. Anthropol.*, 1893.

for the length of the head, a relation which is less clear in the curves for breadth, but still suggested by them. Moreover, since the curves for both measurements fail to indicate that these growth changes have ceased even at the age of twenty-one years, further observations would be of value.

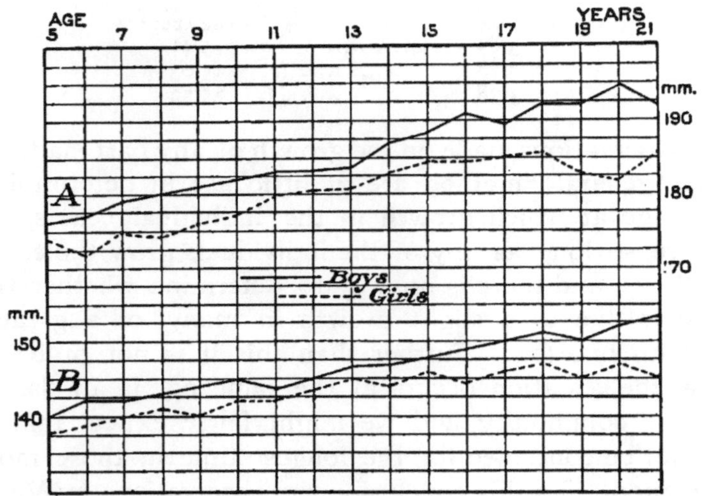

FIG. 18.—Showing the changes in the length and breadth of the head between the ages of five and twenty-one years. A. The curve for length. B. The curve for breadth. (West.)

The records of brain-weight do not, however, clearly indicate increase beyond this age, but in making any comparison it must be borne in mind that the observations by West were on living children developing under fairly favourable conditions, while those collected by Vierordt were from necropsies on the least favoured classes, and also that all the increase in these dimensions of the head is not to be directly referred back to the encephalon, but due allowance should be made for the increase in the thickness of the scalp and bones, and also for the space between the brain and dura.

The brain-weight among non-European races has been but little studied, and nothing is to be gained by attempting to combine the more scattered observations that exist. In the United States, however, during the war of the secession there were made by Dr. S. B. Hunt[1] a series of observations on the brain-weight of negroes and mulattoes as compared with that of native white soldiers. It is probable that in all cases the negro blood in the mulatto was mainly derived from the mother.

TABLE 21*a*.—THE WEIGHT OF THE ENCEPHALON IN AFRICAN NEGROES AND IN MULATTOES. (*Hunt.*)

NO. OF OBSERVATIONS. ALL MALES.	GRADE OF COLOUR.	AVERAGE BRAIN-WEIGHT.	MAXIMUM.	MINIMUM.	DISTRIBUTION OF CASES.						
					>1701	1700 to 1559	1558 to 1418	1417 to 1276	1275 to 1134	1133 to 992	<992
24	White	1475	1814	1254	1	4	11	7	1	—	—
25	12/16	1391	1729	1134	1	—	10	12	2	—	—
47	8/16	1334	1616	1070	—	2	13	19	12	1	—
51	4/16	1318	1673	1091	—	2	10	22	11	6	—
95	2/16	1307	1616	978	—	1	15	50	21	7	1
22	1/16	1279	1432	1134	—	3	10	9	—	—	—
141	Black	1332	1587	1013	—	5	42	51	38	3	—
278[2]	White	1403	1843	964	7	28	99	97	39	7	1

The absolute weights are large both for whites and blacks. But it must be recalled that these subjects

[1] Dr. S. B. Hunt, *Quart. Journ. Psychol. Med.*, New York, 1867.
[2] Compiled from observations by Clendenning, Sims, Tiedemann, and Reid.

were from a more vigorous class of persons than is ordinarily found in the hospitals of cities. Statistically, the results are satisfactory. They show the male negro brain to be less heavy than that of the average white soldier, and that in mulattoes (the father being white) a mixture of less than half white blood gives a brain-weight below that of the pure negro, whereas in those possessing half or more than half white blood, the weight of the brain is above that figure.

Recently Gilchenko [1] has reported on the weight of the brain in several transcaucasian races. The weight was that of the entire encephalon covered with pia, and was taken without drainage. These results are of especial interest as showing the wide variation which occurs among races of the same region and fairly similar in their culture.

TABLE 21*b*.—SHOWING THE WEIGHT OF THE ENCEPHALON IN SEVERAL TRANSCAUCASIAN TRIBES. WEIGHT TAKEN WITH PIA AND WITHOUT DRAINAGE. (*Gilchenko.*)

NO. OF CASES.	RACE.	SEX.	AGE. YEARS.	MEAN STATURE. IN mm.	MEAN WEIGHT. ENCEPHALON.	MAXIMUM.	MINIMUM.
10	Ossétes ...	Males	21–34	—	1470	1541	1306
15	Ingouches	„	18–30	1704	1453	—	—
2	Tcerkesses	„	—	1695	1532	—	—
3	Daghestan	„	—	1650	1340	—	—
12	Armenian	„	16–60	1634	1369	1545	1232
13	Georgian	„	19–65	1669	1350	1530	1183
2	„	Females	25–28	1590	1207	—	—

The older tables, giving the weight of the brain in sixteen Chinese (Crochley-Clapham), and in nine Turcos, Algerian soldiers in the French army (Bischoff), are to be found in Bischoff's *Hirngewicht des Menschen.*

The study of the capacity of the human cranium very

[1] Gilchenko, *L'Anthropologie*, t. iii., No. 5, 1892.

early suggested itself as a means of determining the weight of the brain among those races which were difficult of direct observation. To determine the capacity of a skull requires much art, despite the apparent simplicity of the procedure. For this determination the cranium is usually filled with either shot, water, or some of the varieties of small round seeds, and the amount of this material which it will hold is subsequently weighed or measured. There are more difficulties connected with the operation than would at first sight appear, but leaving them aside, if a given observer in a uniform way determines the capacity of a series of skulls, the figures which he obtains will be fairly comparable among themselves. For example, the table which is given below is based upon observations by Barnard Davis, in which, by a uniform method, the capacity of a large number of skulls was taken, and then by a formula the brain-weight deduced from the cranial capacity.[1]

TABLE 22.—BRAIN-WEIGHTS OF DIFFERENT RACES AS CALCULATED FROM THEIR CRANIAL CAPACITIES. (*Davis.*)

Races.	MALES.				FEMALES.			
	No. of Cases.	Heaviest.	Lightest.	Average.	Average.	Lightest.	Heaviest.	No. of Cases.
European	299	1364—	1212	1340	1180	1099—	1278	94
Oceanic	210	1369—	1192	1293	1185	1139—	1239	95
American	52	1338—	1209	1282	1164	1087—	1263	31
Asiatic	124	1397—	1155	1278	1171	1042—	1276	86
African	53	1316—	1165	1268	1187	1100—	1220	60
Australian	24	1414—	1027	1190	1089	966—	1194	11

These figures are given for both sexes and are divided into six geographical groups. The groups are

[1] Davis, *Journ. of Acad. Nat. Science*, Philadelphia, 1869.

rearranged in the order of the average weight of male brains, which is different from the order in which Davis himself has published them. Supposing his method to have been reliable, we have here a table giving at least the relative brain development in the races compared. The European races head the list and the Australian group is at the foot; but it is plain that if the attempt were made to arrange the intermediate groups according to the degree of their supposed intelligence, they would not stand in the same relation that they now do.

Difficulty also attends the attempt to correlate the weight of the brain with the stature of the several races. While in general, among Europeans, the taller individuals in a given race have the heavier brains, it does not follow that the taller races have the heavier brains, so that a correlation between brain-weight and stature is only applicable within narrow limits. The correlation is closer, apparently, with the mass of the body than with its height. But concerning remote races there exist for the most part only such general facts as that they are lightly or heavily built, and no careful determinations of the weight.

At best a conclusion from bodily development to brain-weight is hazardous. Indeed, the fact that the brain at the seventh year of life has reached almost its full weight and the skull corresponding to it almost its full capacity, while the stature of these same individuals is but two-thirds, and the body-weight only about one-third of what it will be at maturity, makes it readily appreciable that it is impracticable to directly correlate the two growth processes, since they take place at different periods, the increase in stature and weight following in so large a measure the practical completion of growth in the brain. Moreover, Davis and others have pointed out that among all the geographical groups, except the European and the Australian, there are two types of brain, a light and a heavy, these two types being well marked in the Oceanic, Asiatic, and African races, though the average figures here given do not take account of them.

The extreme difference shown in this table between the average weights for the males is (European, 1,340; 1,190, Australian) 150 grammes, while the European female is 160 grammes, less than the male; thus putting the Australian male and European female on the same level as regards gross brain-weight, and thereby suggesting that the inference from brain-weight directly to intelligence is not a happy one. There are several circumstances which here influence the determination of the difference between the sexes. On the one hand, in grouping skulls according to sex by means of osteological characters, those male skulls which have the special characters poorly developed are placed in the female group, and *vice versâ*. Since capacity is not dependent on these characters, the result of this is to put some larger male skulls in the female group, and *vice versâ*, and so to make more equal the figures for the two sexes. At the same time, Davis assumed that the relation between the weight of the brain and the capacity of the skull was the same for both sexes. From the observations of Bischoff,[1] there is reason to think that in the female the brain more nearly fills the cranial cavity by from 3 to 4 per cent. Introducing this correction into the table would alter the result materially by raising the average weight for the female. But, granting all corrections, the most striking feature of the table, the high average brain-weight of the European male, would not be materially affected.

[1] Bischoff, *Hirngewicht des Menschen*, 1880.

In the preceding paragraphs it has been the value of the relative figures within a single series of observations that has been under discussion, but when from the cranial capacity the attempt is made to deduce the size of an individual brain, the usual difficulties appear. The authors are fairly well agreed that the relations of the brain to the cranial cavity vary with age, sex, and cause of death. The brain appears more nearly to fill the cranial cavity in the young than in the adult ; in old age there is an increasing diminution in both weight and volume. In females it more nearly fills the cranial cavity than in the males, and direct observation shows that its volume may be less than that of the cranial cavity by from 7 to 33 per cent. of the latter. Manouvrier found that sex being neglected, the brains which he[1] examined were about 16 per cent. less in volume than the cranial cavity ; but the observations were made on dried skulls, and these compared with fresh brains, no correction having been made for the shrinkage of the skulls. This shrinkage, due to drying, would appear from certain figures published by Bischoff to reduce the cranial capacity from 1 to 2 per cent. Despite these data, when the cranial capacity of a single skull is alone given, it is possible only with much uncertainty to deduce the size of the particular brain which filled it. On the capacity of the skull according to age, a few observations have been recorded by Broca. At any given time it must always have a capacity somewhat greater than the volume of the brain, but whether the relations of growth are such that an enlargement of the skull is followed by the enlargement of the brain, or whether they grow simultaneously, has yet to be determined.

[1] Manouvrier, *Sur l'interpretation de la quantité dans l'Encéphale*, &c., Paris, 1885.

CHAPTER VI.

VARIATIONS IN BRAIN-WEIGHT.

Cranial capacity in various races—Giants and dwarfs—Brain-weight in animals—Relation of the weight of the brain to the weight of the body—Nomenclature of the encephalon—Microcephalics—Weight of microcephalic brains according to sex—Argument for the correlation of weight with intelligence—Brain-weight and cranial capacity of eminent men—Cranial capacity of murderers—Macrocephalic brains—Physiological superiority of mean forms—Brain-weight in the insane—Proportional development—Relation of brain-weight to mental disease—Pathological changes affecting brain-weight in the insane—Résumé.

HAVING presented some of the observations relating to the normal size of the brain and the history of the steps by which that size is attained, we naturally turn to the study of the extreme variations in size, because difference in size has been made the basis of many current views. The measurements of the capacity of skulls have shown that the different races of men differ widely with regard to the probable size and weight of their brains; and that in general those races which are inferior both in stature and in body-weight, have, at the same time, a small cranial capacity. While the heaviest brain-weights belong to the European races and the lightest to the Australians, thus giving a moderately wide difference in the weight of the brain corresponding to a wide difference in culture, yet it is quite impossible even in such a condensed series to harmonise the intermediate groups with the theory that brain-weight and culture, as we measure it, are closely correlated.

If attention is confined to the European peoples with whom we are most familiar, it is found that occasionally there occur individuals noticeable for their deviation from the general average by reason of their weight or stature or both. These persons of unusual stature are the giants and dwarfs. It is customary, among English authors, to consider individuals more than seven feet (2,133 mm.) tall as giants, and those under four feet (1,219 mm.) as dwarfs. Examination shows that the giants owe their excess in stature either to an abnormal length of the lower limb, or to an excessive development in which the proportions of the normal body are fairly well maintained. As a rule, however, the head in giants is disproportionately small, and also, as a rule, they are neither mentally nor physically possessed of unusual powers. In fact, their general health tends to be poor, and they are short-lived. These facts are still without a full explanation, though it is possible that in one group a nervous disease (acromegaly) is responsible for the result.[1]

On the other hand, the dwarfs have more vigour, and tend to be nearly normally developed, with the exception of one well-marked group among them, in which the excessive shortness of the lower limbs accounts for the small stature. In these deformed dwarfs a rachitic condition is often a prominent feature. In general in this group the trunk is well developed, the head proportionately large, and the individuals active, both mentally and physically. We have much to learn concerning such exceptional persons, but there are no observations

[1] Dana, "Giants and Gigantism," *Scribner's Magazine*, vol. xvii., No. 2, Feb., 1895.

on the histological structure of their bodies which would enable us to interpret the differences between them in terms of their cellular constitution. There is some reason to think, however, that in dwarfs the cell elements are of small size. In every way it would appear that the physiological condition of the dwarf is superior to that of the giant, and that he represents the more efficient organism. These facts have the following bearing: In general the presumptive brain-weight among giants is greater than among the dwarfs, yet to put the case most conservatively, the giants are not recognised in any sense a mentally superior group. Hence under these conditions the larger brain is not correlated with the greater intelligence.

This relation of brain to body may be considered from another point of view. It has always been assumed that some light would be obtained from comparative anatomy, and to that end many observations have been made in various animals on the weight and size of the brain, and its relation to the weight of the body. It appears that man is surpassed in the gross weight of the encephalon only by the elephant, some of the whales, and the recently extinct Stellar's sea-cow. When the weight of the brain is compared with the weight of the fully grown body, its proportional value is found to be higher in man than in most animals, although some of the lower monkeys, small rodents, and some birds, have a proportional brain-weight which is greater than that in man. Attention has already been called to the fact that this proportional value of the brain is greatest at birth, and diminishes throughout the growing period. The same is probably true of all vertebrates. As an illustration of this point a table on the relation between the weight of the brain and the weight of the body in several species of dogs is here quoted from Wilder.[1]

TABLE 23.—THE WEIGHT OF THE BRAIN AND OF THE BODY, AND THE RATIO OF THE TWO, IN DOGS OF DIFFERENT SIZES. (*Wilder.*)

CASE.	VARIETY.	AGE.	BODY-WEIGHT.	BRAIN-WEIGHT.	RATIO.
1	Pomeranian .	54 hours	132 gms.	8 gms.	1:c60
2	English Terrier (small)	6 months	1320 „	38 „	1:028
3	English Terrier (large)	3·5 years	5300 „	69 „	1:013
4	Newfoundland	Adult	38345 „	120 „	1:003

There are two factors at work to bring about the different ratios here given. In the first place, the brain at birth has more nearly completed its growth than has the body ; young dogs, like the first two mentioned, have, in common with all other young vertebrates, the brain disproportionately large ; and in the second place, the large varieties grow for a longer time. Cases 3 and 4 in this table may be cited, as showing the difficulty of making a correlation between brain-weight and intelligence. The large English terrier, 3·5 years old, has approximately only one-seventh of the body-weight, and but one-half of the brain-weight, of the Newfoundland. The ratios are correspondingly unlike, yet certainly both of them belong to the breeds counted as intelligent, and it should be safe to say that the difference between the respective ratios of their brain and body-weights is out of all proportion to the presumptive difference in their mental powers.

[1] Wilder, *Rep. Am. Assoc. Advancement of Science*, 1873.

On comparing the two sexes in the human race, the brain, compared with the entire body, is found to be proportionally heavier in the female than in the male. In this instance the only logical interpretation is that the body is somewhat less developed in the female than in the male, and this we know to be the case from the records on the growth of the muscular system. For this reason no effort will be made to restate the various observations on the proportional development of the brain in the two sexes, or the number of grammes of body substance present for each gramme of brain. In such arguments it is assumed that the smaller the mass of the body over which a unit of brain substance presides, the more efficiently will the body be controlled, and the more intelligent will be the individual. This view is not well grounded. While the large absolute weight of the brain in man as compared with other animals still demands interpretation, it is evident from the foregoing facts that conclusions based upon the proportional development of the brain may be very easily misleading, because the variation in that proportion depends upon differences in body-weight. In reality there is here an extremely complex problem, and although in the estimation of the probable intelligence of any animal both the absolute and relative size of the brain are factors, yet they are but two factors among a large number, and therefore their importance can only be determined after further analysis and comparison.[1]

Returning to the principal question before us, we have to inquire whether from any of the physical characters of the brain the intelligence of the individual can be inferred. By way of introduction it will be necessary to

[1] For a general expression of the percentage weight of the brain in terms of the entire body, consult the tables by Vierordt, given in Chapter III., pp. 69 and 70.

examine the entire range of brain-weight for the adult. This is given in Table 24, where the brains are arranged in five classes, according to size, namely, macrocephalic, large, medium, small, and microcephalic.

TABLE 24.—THE NOMENCLATURE OF THE ENCEPHALON ACCORDING TO WEIGHT. WEIGHT IN GRAMMES. (*Topinard.*)

CLASSES.	MALES.	FEMALES.
Macrocephalic ...	From 1925–1701	From 1743–1501
Large	,, 1700–1451	,, 1500–1351
Medium	,, 1450–1251	,, 1350–1151
Small	,, 1250–1001	,, 1150–901
Microcephalic ...	,, 1000–300	,, 900–283

Excluded from these tables, of course, are the cases of monsters born with comparatively little functional brain substance, but which have continued to live for months. Directly continuous with these excluded cases is the group of defective individuals with heads ranging from extremely small to small, and designated as microcephalics. The weight below which the brain is designated as microcephalic is 1,000 grammes in males, 900 grammes in females. The best list of these microcephalic cases has been compiled by Marchand, and contains some fifty entries, including male brains [1] weighing up to 1,015 grammes, and female brains up to 924 grammes, but it is unnecessary to repeat it here *in extenso*. Examination of the table shows that among children the smallest brain-weight is reported by Calori, the case being that of a female child 9 months of age, 53 cm. high, and possessed of a brain weighing 69·3

[1] Marchand, *Acta. d. Kaiserl. Carol. Deutsch. Akad. der Naturforscher*, Halle, 1890.

grammes. Among the microcephalics, in which the brain may be considered as having attained approximately its full weight—*i.e.*, those more than seven years old—we have the case of Helene Becker cited by Bischoff. This girl died at seven years of age with a stature of 78 cm., and a brain-weight of 219 grammes.

If from the fifty cases cited by Marchand children less than ten years of age are excluded, there remain twelve cases in which the brain-weight was less than 500 grammes, fourteen in which it ranged between 500 and 800 grammes, and in eleven cases between 800–1,015 grammes. In these instances the brain is not only of the small size indicated by the enclosing skull, but at the same time the skull is apt to be unusually thick, and either the brain does not fill it so completely as normally, or the enlargement of the cerebral ventricles by fluid causes the actual weight of the nerve substance to be smaller; so that when the volume of the brain is deduced from its weight on the assumption that it was normal in its proportions, it is in some instances found to be much smaller than the cranial cavity. Furthermore, even this mass often contains a large proportion of undeveloped or degenerate tissue, and the encephalon may thus be of even less functional value than its small size would suggest. It thus appears that the essential anatomical feature in microcephalism is not so much the small size of the brain as its deficient construction, of which this small size can usually be taken as an index. On the other hand, that small size and great faultiness in construction are not necessarily connected is suggested by Marchand, who has pointed out that the microcephalics are by no means necessarily aphasic, and that one Antonia Grandoni, a woman living to be forty-one years of age, with a stature of 132 cm., and a brain-weight of 289 grammes, both sang and spoke readily, while other microcephalics have possessed more command over spoken language, and still others have made use of deaf-mute signs as a means of communication. The senses are often acute and the motor activity good, and in such cases, of course, there is a rudimentary intellectual life. In other words, this class represents individuals with diminished intelligence, but even a very small brain-weight, as is shown in the case just cited, is not incompatible with the performance of simple mental processes; at the same time it cannot be assumed that these individuals were sufficiently well organised to make life possible for them if they had not been specially protected. In every group of this table of Marchand the female brain-weight is distinctly less than that of the male. In tabular form the facts appear as follows:—

TABLE 25.—THE WEIGHT OF THE BRAIN IN MICROCEPHALICS.
(*Condensed from Marchand.*)

In each of the three groups taken the average weight for the females is less than that for the males.

GROUP.	241–500 GRMS.	501–800 GRMS.	801–1015 GRMS.
Males } Average Weight	349	651	954
Females } Average Weight	299	621	912

Since in these groups the two sexes are presumptively on a par mentally, having their intellectual processes very much reduced, it is certainly curious that this common mental level should still be accompanied by a higher brain-weight in the males than in the females. The usual argument assumes that the simplest psychic powers are, as indicated in the table, correlated with a minimal mass of brain substance which, other things being equal, should be the same in both sexes. It further requires that when the brain-weight increases there should be an increased mental capacity, and finally, since it is found that the average European male has a brain-weight about 120 grammes in excess of that for the female, it is therefore concluded that the male brain has superior powers, of which the 120 grammes can serve as a measure. Since among microcephalics there is still a distinction in brain-weight according to sex, the current explanation must be considered as incomplete.

On passing from the microcephalics to other groups in which the weight of the brain is greater, we come into the company of those individuals who are able to make a fairly successful struggle for existence, a capacity which is taken as one measure of the intelligence. It is perfectly plain, however, that a wild life under clement conditions can be met by a far simpler mental organisation than is needed to face town life in the temperate zone, and hence the capacity to lead an independent existence is but a poor standard.

The observations recorded in the preceding chapters compare average brain-weights (small, medium, and large being taken together) with one another, and show several conditions which influence this average, but as they stand the figures furnish no basis for correlating brain-weight with brain functions. Attempts have been made, however, to get some light on this from the comparison of the different classes in the community. Take, for example, the brain-weights of eminent men, arranged according to age with omission of certain mythical records like those for Tourgenieff (2,012 grammes), Cromwell (2,231 grammes), and Byron (2,238 grammes).

TABLE 26.—GIVING THE BRAIN-WEIGHTS OF CERTAIN EMINENT MEN, COMPILED FROM RECORDS BY MARSHALL AND MANOUVRIER.[1]

AGE.	ENCEPHALIC WEIGHT.	EMINENT MEN.
39	1457	Skobeleff, Russian General.
40	1238	G. Harless, Physiologist.
43	1294	Gambetta, Statesman.
45	1403	Assezat, Political writer.
45	1516	Chauncey Wright, Mathematician.
49	1468	Asseline, Political writer.
49	1409	J. Huber, Philosopher.
5 (?)	1312	Seizel, Sculptor.
50	1378	Coudereau, Physician.
52	1358	Hermann, Philologist.
52	1499	Fuchs, Pathologist.
53	1644	Thackeray, Novelist.
54	1520	De Morny, Statesman.
54	1629	Goodsir, Anatomist.
55	1520	Derichlet, Mathematician.
56	1503	Schleich, Writer.
56	1485	Broca, Anthropologist.
57	1559	Spurzheim, Phrenologist.
57	1250	v. Lasualx, Physician.
59	1436	Dupuytren, Surgeon.
60	1533	J. Simpson, Physician.
60	1488	Pfeufer, Physician.
62	1398	Bertillon, Anthropologist.
62(?)	1415	Melchior Mayer, Poet.
63	1449	Lamarque, General.
63	1332	J. Hughes Bennett, Physician.
63	1830	G. Cuvier, Naturalist.
64	1785	Abercrombie, Physician.
65	1498	De Morgan, Mathematician.
66	1512	Agassiz, Naturalist.
67	1502	Chalmers, Preacher.
70	1352	Liebig, Chemist.
70	1516	Daniel Webster, Statesman.
71	1207	Döllinger, Anatomist.
71	1349	Fallmerayer, Historian.
71	1390	Whewell, Philosopher.
73	1590	Hermann, Economist.

[1] The entries in this table have been in part revised. Different methods have of course been employed in determining the several weights.

AGE.	ENCEPHALIC WEIGHT.	EMINENT MEN.
75	1410	Grote, Historian.
77	1226	Hausemann, Mineralogist.
78	1492	Gauss, Mathematician.
79	1254	Tiedemann, Anatomist.
79	1403	Babbage, Mathematician.
79	1452	Ch. H. Bischoff, Physician.
80	1290	Grant, Anatomist.
82	1516	Campbell, Lord Chancellor.

The application of the term "eminent" to men whose names stand in the preceding table, is, in some measure, a concession to custom. While it is easy to select from this list those recognised as persons of first-rate ability, there are others whose accomplishments would by no means warrant such recognition. The list, therefore, contains not only the brain-weights of a number of persons unquestionably remarkable, but also those of other persons whose title to our regard is merely the successful pursuit of some learned profession. This latter group, to be sure, is the one concerning which more information is most needed.

These data have been in part utilised by Manouvrier, who has published the following table, based on the observations by Broca and by himself, in which a comparison is attempted between the brain-weight of eminent as compared with that of ordinary persons. The range in the weight of the encephala is represented at the left, Table 27, and then in each group of individuals the percentage of cases occurring within the given limits is noted. The weights around which the greatest percentage of the cases centre are those of most significance.

TABLE 27.—THE BRAIN-WEIGHTS OCCURRING AMONG EMINENT MEN, AS COMPARED WITH THOSE OF PARISIANS OF DIFFERENT STATURES. (*Manouvrier.*)

The figures express the number of brains in each hundred which would fall within the limits of weight opposite to which the entries stand.

WEIGHT OF THE ENCEPHALON.	PARISIANS OF BROCA, ADULT.	PARISIANS OF TALL STATURE.	EMINENT MEN.		
In Grammes.	168 cm.	171–185 cm.	1st Series.	2nd Series.	Series 1 and 2 combined.
900–1000	0·6	—	—	—	—
1001–1100	0·6	—	—	—	—
1101–1200	7·1	3·5	—	2·9	1·2
1201–1300	23·3	15·5	11·1	2·9	7·5
1301–1400	31·5	27·5	17·8	17·2	17·5
1401–1500	23·8	34·6	33·3	48·5	40·0
1501–1600	9·6	15·5	24·5	22·8	23·8
1601–1700	3·5	3·4	2·2	5·7	3·3
1701–1800	—	—	2·2	—	1·3
1801–1900	—	—	2·2	—	1·3
1901–2000	—	—	—	—	—
2001 and more[1]	—	—	6·7	—	3·8
Total	100	100	100	100	100

The first series of ordinary Parisian men shows, as we should expect, the maximum frequency of 31·5 per cent. of cases between 1,300–1,400 grammes. Examining the Parisians of tall stature, we find the point of maximum frequency amounting to 34·6 per cent., and raised to between 1,400–1,500 grammes. On comparing the eminent men with this latter group, it is found that the maximum frequency falls also between 1,400–1,500 grammes, but there is this difference, that among eminent men there is a decidedly larger number of cases between 1,500–1,600 grammes, and also beyond

[1] These observations have been excluded from Table 26.

1,700 grammes a small percentage of the cases not at all represented in the records for ordinary persons. The whole group of eminent men is therefore moved slightly upward in the scale of weight.

Before offering any further comment on this table, I shall present another in which the cranial capacities of modern Parisians, murderers, and eminent men are compared. The purpose of this table is similar to that of the foregoing, and it differs from it only in requiring the additional assumption that the brain-weights are related in nearly the same manner as are the cranial capacities.

TABLE 28.—SHOWING THE CRANIAL CAPACITIES OCCURRING AMONG EMINENT MEN AND MURDERERS AS COMPARED WITH THOSE IN TWO GROUPS OF MODERN PARISIANS, ONE MEASURED BY BROCA, THE OTHER BY MANOUVRIER. THE CONSTRUCTION OF THE TABLE IS SIMILAR TO THAT OF TABLE 27. (*Manouvrier.*)

CRANIAL CAPACITY.	MODERN PARISIANS.		MURDERERS.	EMINENT MEN.
In Cubic Centimeters.	70 (Broca).	110 (L.M.).	45 (L.M.).	35 (L.M.).
From 1200–1300	—	1·8	—	—
,, 1301–1400	10·4	10·0	8·9	2·9
,, 1401–1500	14·3	21·8	17·8	2·9
,, 1501–1600	46·7	30·0	33·3	17·2
,, 1601–1700	16·9	17·4	17·8	34·2
,, 1701–1800	6·5	14·5	13·3	34·2
,, 1801–1900	5·2	4·5	6·7	8·6
Above 1900		—	2·2	—
Total	100·0	100·0	100·0	100·0
Average capacity	1560 c.c.	1560 c.c.	1571 c.c.	1665 c.c.

From the examination of Table 27 it appears that the greater brain-weights are more frequent among the eminent men, even when these latter are compared with the group having the greatest stature; at the same time there is no doubt that the tall persons here taken as a standard belonged to a less favoured social class. There is every reason to think that the favoured classes in the community will have a brain-weight superior to that of the class from which all our ordinary averages have been derived, and the comparison therefore fails to be exact, so that although there is no doubt about the facts which have been given, their interpretation must be postponed. To Table 28, based on cranial capacities of eminent men, the same argument applies. In this table there is also introduced a series of observations on the cranial capacity of murderers. This series is of interest in connection with the attempt to determine whether the criminal has bodily peculiarities which mark him. As will be seen, the criminals are very close to the modern Parisian in their cranial capacity, and even somewhat surpass them, though falling distinctly below the distinguished men with whom they are compared. It would appear from this that the criminals in question had been mainly derived from the same class that furnished records with which they are here compared. There exist no measurements either on the brains or skulls of eminent women, and the differences due to sex, which have been so much discussed in other groups, are therefore entirely wanting in this one. From these facts it is plain that the brains of eminent men are the heaviest, and their skulls the most capacious, but we know little of the brain-weight and cranial capacity of their neighbours—men successful in business and professional life—who, though not distinguished, grew up and lived under like conditions. As far as it goes, the evidence only shows that the members of the less fortunate social classes have the smaller brain-weight. An emphasis of this point is warranted because the central nervous system, whatever its natural perfection, must be extremely responsive to surrounding social conditions, and thus growth processes in it be modifiable in no small degree, hence the conditions which social status implies are probably important.

The group of very large brains yet remains to be considered. The weight recorded for the brain of Oliver Cromwell, 2,231 grammes; for Byron, 2,238 grammes; and for Turgenieff, 2,012 grammes, would have given these persons heads which could hardly have escaped description or portraiture, by which the post-mortem records might be corroborated. But there is a complete absence of such collateral evidence, save in the last instance, besides much evidence in the case of the first two, positively opposed to the correctness of the recorded weights; while the last case has been reported without detail. In any long series of observations on brain-weight there are always found a few macrocephalic brains. For example, in Bischoff's tables, giving the weight of 559 male brains, there are 59 cases above 1,501 grammes, and 18 cases above 1,601 grammes, and one, a mechanic, with a brain-weight of 1,925 grammes. It happens that among these 59 cases given by Bischoff, there is one learned man, Dr. Hermann of Munich, 13 criminals of various sorts, and next to the mechanic just mentioned, comes a day labourer, with a brain-weight of 1,770 grammes. These conditions would be found repeated in other long series. It follows, therefore, that within a given series there is no evidence that the persons possessing the very large or macrocephalic brains, form the more intelligent fraction of the group. In this group, also, the female brains of the macrocephalic type, as shown in Table 24, never attain so great a weight as in the case of the males, although statistically they are just as abnormal. Pathological hyperplasia of brain substance suggests itself to explain some of these cases. The brain is sometimes described as expanding on the removal of the skull cap, suggesting that it had been compressed during life, and in some of these cases part of the excessive growth is due to an increase in the supporting tissues, but further data are wanting.

From this survey we finally come back to the very general idea that in the case of the brain, as among all organisms and organs called similar, there may be wide individual variations, and as it is midway between the extremes that the greatest number of cases occur, therefore we must consider the most numerous forms to be those best adapted to surrounding conditions, and consequently also look to them for the most perfect physiological reactions. If such reasoning is correct, we should hardly expect to find the best intelligence usually associated with excessive brain-weight. At the same time, as soon as small or macrocephalic brains are examined, deficiencies in the intelligence appear, and, although there is no exact correspondence between the two forms of loss, yet at the lower end of the series the connection between vanishing intelligence and minimal brain-weight is close.

In the foregoing paragraphs the variations of the brain-weight among sane persons have alone been described ; it remains for us to examine these relations among the insane, and from the facts in this quarter select such as will bear upon the problem. The great series of records by Dr. Boyd contains observations on the insane as well as the sane. There were 400 observations on insane males and 325 on insane females. Tabulated in the same manner as those for the sane, they stand as follows :—

TABLE 29.—SHOWING THE WEIGHT OF THE ENCEPHALON AND ITS SUBDIVISIONS IN INSANE PERSONS, THE RECORDS BEING ARRANGED ACCORDING TO SEX, AGE, AND STATURE. (*From Marshall's tables based on Boyd's records.*)

a indicates that a record considered according to age is too large; s indicates that a record considered according to stature is too large.

INSANE.									
MALES. No. of Cases = 400.					FEMALES. No. of Cases = 325.				
Ages.	Encephalon.	Cerebrum.	Cerebellum.	Stem.	Stem.	Cerebellum.	Cerebrum.	Encephalon.	Ages.
Stature 175 cm. *and upwards.*					*Stature* 163 cm. *and upwards.*				
20–40	1378	1192	156	30	28	136	1056	1220	20–40
41–70	1354	1170	154	30	28	134	1053	1215	41–70
71–90	1333	1158	146	29	28	136 a	1076 a	1240 a	71–90
Stature 172—167 cm.					*Stature* 160—155 cm.				
20–40	1363	1186	149	28	28	134	1027	1189	20–40
41–70	1305	1129	148	28	27	135 sa	1054 sa	1216 sa	41–70
71–90	1305	1135 a	142	28	28 a	135 a	1008	1171	71–90
Stature 164 cm. *and under.*					*Stature* 152 cm. *and under.*				
20–40	1299	1127	144	28	28	128	986	1141	20–40
41–70	1285	1119	139	28	28 sa	129 a	1036 a	1194 a	41–70
71–90	1216	1047	139	30 s a	27	123	985	1135	71–90

When these figures are compared with those for the sane, in Table 13, it appears that in the insane males the encephala in two groups are heavier than in the sane. The mean weight for the insane males, however, is slightly less than for the sane (15 grammes). When the females are compared in the same way, it is found that in the insane five groups show the encephalon heavier than in the sane, and the mean weight in both classes is the same. Since in the insane the wasting forms of mental disease are proportionally much more numerous among the males than the females, the smaller difference in weight according to sex in the insane becomes explicable.

It appears from this that the insane are not a class with a characteristic brain-weight. The absolute increase in the weight of the stem is noticeable in both sexes, while the cerebellum undergoes the most decided increase in the case of the male, and both peculiarities are most evident in tall persons. When among the insane we compare the two sexes, it is plain that the difference in the proportional development of the hemispheres is slightly less than among the sane. This is probably due to the wasting processes which tend to reduce the weight of the male encephalon as a whole, affecting especially the cerebral hemispheres; but when the averages for stature independent of age, and for age independent of stature are taken, they are found strikingly similar under both conditions. The figures are so uniform that a single record of the proportion—one for age and one for stature—will suffice.

TABLE 30.—SHOWING IN THE UPPER LINE THE PERCENTAGE WEIGHT OF THE SUBDIVISIONS OF THE ENCEPHALON BETWEEN 20–40 YEARS, WITHOUT REGARD TO STATURE. IN THE SECOND LINE THE SAME FOR THE GROUPS OF TALLEST INDIVIDUALS, WITHOUT REGARD TO AGE. THE DIFFERENCE IN THE PERCENTAGE OF THE CEREBRUM AT OTHER AGES OR OTHER STATURES IS AT MOST ·3 PER CENT. (*Based on Table* 29.)

AGE.	ENCEPHALON.	CEREBRUM	CEREBELLUM.	STEM.	STEM.	CEREBELLUM.	CEREBRUM.	ENCEPHALON.	AGE.
20–40	100	86·7	11·1	2·2	2·3	11·2	86·4	100	20–40
Stature. 175 cm. and upwards	100	86·6	11·2	2·2	2·2	11·1	86·6	100	Stature. 163 cm. and upwards

The proportion of the cerebral hemispheres is less than in the sane, and as a consequence the percentages for the stem and cerebellum are higher, and this is due not only to the diminution in the absolute weight of the cerebrum, but also to the increase in that of the cerebellum and stem. The insane, therefore, present encephala that are very slightly less than the normal in absolute weight, and in proportional development show a slightly smaller figure for the cerebral hemispheres, while the difference between the two sexes is less marked in them than in the sane.

The different groups of the insane have been examined by several investigators with a view to determining any possible connection between encephalic weight and the several forms of mental diseases. With the insane are usually reckoned the congenital idiots, and those whose mental defect is due to arrest of growth during the period of development. This group is nearly identical with that of the microcephalics, and as that has been already discussed, it may here be omitted. It is then to be determined whether the fully developed encephalon ranging between small and large is, by virtue of variations in its size, subject to different forms of mental disease. The accompanying Table (31) is based on observations made by Boyd at the Somerset County Lunatic Asylum, England, so that the individuals examined were drawn from a rural as contrasted with the urban population which furnished the cases in the table for the sane; nevertheless the two sets of observations are fairly comparable. As will be seen, it has been made up in a different manner from those which have previously been presented, the cases being classed according to the form of mental disease. For the males they are arranged in series, beginning with the forms in which the encephalic weight was found to be greatest, and placing the others in regular order below. As a result, the three forms of mental disease in which wasting of the brain takes place (the last three entries in the table) are the ones which stand lowest in the series. The percentage values of the different divisions of the encephalon are given for each form, and that for the cerebrum is found to be less in those diseases in which the encephalic weight is less. When the groups for the females are treated in the same way, the average encephalic weight and the percentage value of the cerebrum are both less in the class of wasting diseases than in the other class, but the encephalic weight does not regularly decrease from the first to last group.

TABLE 31.—SHOWING THE WEIGHT OF THE ENCEPHALON AND ITS SUBDIVISIONS IN VARIOUS GROUPS OF THE INSANE. THE PERCENTAGE VALUES ALONE ARE GIVEN. AGE AND STATURE NEGLECTED. (*From Boyd's records.*)

INSANE.										
MALES.						FEMALES.				
Total No. of Cases.	Mental Disease.	Encephalon.	Cerebrum.	Cerebellum.	Stem.	Stem.	Cerebellum.	Cerebrum.	Encephalon.	Total No. of Cases.
		gms.							gms.	
108	Mania... ...	1393	87·1	10·7	2·2	2·5	11·1	86·3	1227	107
30	Recurrent Mania	1383	87·0	11·0	2·0	2·5	10·5	86·9	1238	33
52	Melancholia	1335	86·8	11·2	2·0	2·5	10·7	86·8	1261	68
89	Epilepsy ...	1310	87·2	10·6	2·1	2·5	10·9	86·5	1216	60
279	Average =	—	87·0	10·8	2·1	2·5	10·8	86·6	—	268
49	Dementia ...	1307	86·5	11·2	2·3	2·6	11·7	85·7	1188	61
122	Gen. Paralysis	1304	85·9	11·9	2·2	2·8	11·7	85·4	1162	30
29	Senile Dementia	1259	86·8	10·9	2·2	2·5	11·7	85·8	1226	12
200	Average =	—	86·1	11·6	2·2	2·6	11·7	85·6	—	103

In the forms of mania are found the greatest encephalic weights, while the lowest occur in the forms of general paralysis and dementia. The pathological processes which in the latter groups account for decrease in weight are generally recognised, whereas in the manias the overfilling of the encephalon with blood may in part account for the larger figures obtained. Whether there is any significance to be attached to the weights given for melancholia and epilepsy cannot be determined, and it would also be unwise to draw inferences from the relative weights of the subdivisions of the encephalon in the different forms, since if these were really significant the two sexes would be expected to give similar figures; but such is not the case. Since here, as elsewhere, there appears in encephalic weight the characteristic difference between the two sexes, it might be inferred that the smaller weight for both sexes in the diseases which are associated most clearly with the loss of mental power, is an expression of the breaking down of those structures in the brain on which the mental life anatomically depends, and there is some pathological evidence to support such a view. The original records from which this table has been condensed give the cases under each division, arranged according to age, but not grouped according to stature; and hence the figures in these partial tables do not exhibit the same regularity as those in the tables for the sane, but the number of cases entered in each decennial period is not large enough to warrant the expectation that the law of age would come out clearly, especially as the classification according to stature is lacking. Taking, then, the insane altogether, we find them, as far as their encephalic weight is concerned, very similar to sane persons, and the differences which appear in this table may be looked upon in a large measure as the result of disease, and therefore do not warrant the idea that moderate variation in the weight of the brain is a predisposing cause to any particular form of brain affection.

Bringing together the facts which have been included in this chapter, it is plain that they contribute mainly to a healthy scepticism concerning the current interpretations of brain-weight. The most hopeful method of investigation was the comparison of different classes in the community, the classes being based on intellectual performance. This has proved unsatisfactory, for the reason that the individuals forming the several classes have developed under different social conditions. It is clear, therefore, that the differences in weight are not to be associated with the difference in intellectual capacity, until the possible influence of social conditions has been excluded. The best comparison is that between the criminals and the ordinary persons, since the conditions were probably nearly similar in the two cases. It is certainly very easy to indicate the point where the mental capacities diminish because of the very small size of the encephalon, but it must be remembered that the very small encephala are never simply miniatures of healthy brains, but are always more or less organically defective; and our problem is thereby complicated even in this case, where the relation between size and intelligence seems so evident. It is therefore desirable to discuss this question from the standpoint of the structural elements of the brain, in order to see whether some of the confusing results which come from regarding it *en masse*, may not be cleared up when the variations in the structural units are given a proper value.

CHAPTER VII.

THE NERVE ELEMENTS.

Histological basis—First form of central system—Description of cell—Human types—Chemical characters—Various shapes—Outgrowths—Construction of cell—Changes during development—Volume of neuroblast—Changes in the size of the elements with age—Nerve fibres—The medullary sheath—Size of young cell-bodies and fibres—Calculation of the average size of nerve elements and of their number in man.

IN this chapter the size, shape, number, and growth of the nerve elements will be considered. The central system is a mass of such elements embedded in the structures which support and nourish them, and the changes observed in the whole mass must be therefore the sum of the changes taking place in all the constituents, but since in the matter of enlargement the subsidiary structures play but a minor part, attention will be given principally to the nerve elements.

At an early stage in any developing mammal the central nervous system is represented by a tube, the walls of which are formed by ectodermal cells, and by two strands of similar cells coextensive with the tube and lying on either side of it. These latter give rise to the ganglionic portion of the nervous system, which it is convenient to distinguish from the medullary portion, formed by the tube alone. By flexures of this tube, and by variation in the thickness of its walls due to growth, the complex nervous system of the adult is finally elaborated. At first the walls are formed by a thin layer one or two cells in thickness, then there appear next to the central canal, germinal cells which rapidly divide. A portion of the cells thus formed become young nerve cells—neuroblasts—another portion build up the framework—the neurospongium—in which the nerve elements lie. With these changes the cells

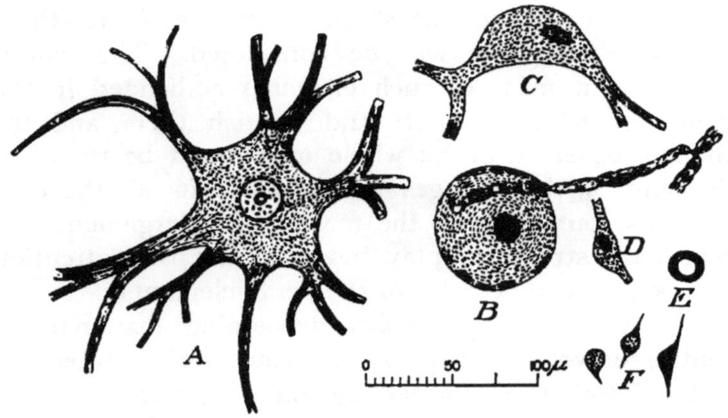

FIG. 19.—A group of human nerve cells drawn to scale, × 200 diameters. *A*, cell from ventral horn of spinal cord; *B*, cell from the spinal ganglion of dorsal root, with its nerve process; *C*, cell from the column of Clarke; *D*, "Solitary" cell from the dorsal horn of the spinal cord; *E*, cross-section of a large nerve fibre; *F*, granule cells from the cortex of the cerebellum. (Modified from Waller, *Human Physiology*, 1893.)

increase in number, then in size, and their form becomes more complicated, while at the same time they separate more widely from one another. To understand these processes, it will be of advantage to describe one or two of the commoner forms of nerve cells after they have reached maturity, returning later to follow them from their first to their final shape. Formerly the terms employed were *nerve cell* and *nerve fibre*, with the implication that these two structures were genetically distinct, but to-day the view is held that the fibre is the outgrowth of the cell, and for that reason I shall follow Schäfer in speaking of the whole element, including all the outgrowths, as a nerve cell ; and then for convenience distinguish between the cell-body and its prolongations.

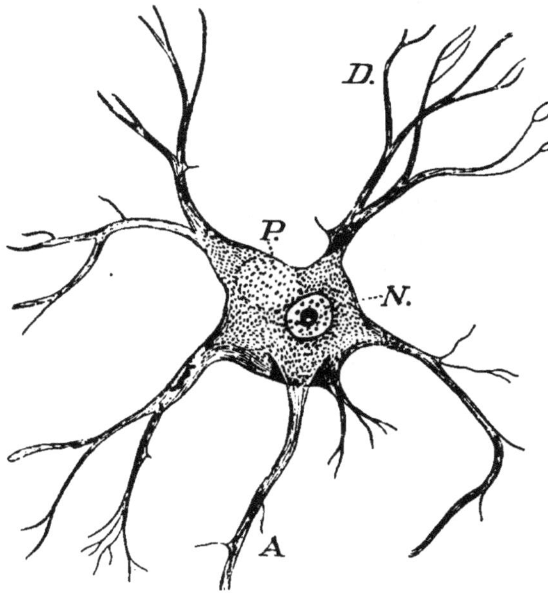

FIG. 20.—Isolated body of a large cell from the ventral horn of the spinal cord. Human, × 200 diameters. (Obersteiner.) *A*, neuron ; *D*, dendrons ; *N*, nucleus with enclosures ; *P*, pigment spot.

Fig. 19 shows a series of cells from the nervous system of man.

A detailed description of the various shapes and sizes of such cells would not be in place, but several types may be mentioned. The largest cells, which are polyhedral, are found in the ventral horns of the spinal cord and in the morphological continuation of these horns into the brain-stem. Next in size come the large pyramidal cells in the cerebral cortex, the cells of Purkinje in the cerebellar cortex and the great ovoid cells in the spinal ganglia. Besides these there are to be found a host of smaller forms everywhere, except in the spinal ganglia, where very small cells do not occur. Fig. 20 represents a fully-grown large nerve cell from the ventral horn of the spinal cord of man. This may be taken as the basis for a more extended description.

The description of the nerve cell begins with the nucleus, since that structure is the first to be clearly developed. In the early stages the nucleus is ovoid and contains a chromatic or deeply staining substance in small masses. Representing at the start almost the entire cell, the nucleus becomes later surrounded by what is really an enormous quantity of cytoplasm, and also forms within itself one, or even two, inclusions. The larger of these, the nucleolus, is deeply stainable, and this in turn may develop the second inclusion, the nucleololus, much more tiny than itself, and noticeable only in the largest cells. It is the cytoplasm, of course, which, according to the locality, may be pyramidal or ovoid in form. Perhaps the most noticeable substance in the cytoplasm is the pigment. As a rule this is not present at birth, but appears later, and increases with age.

A stainable substance is also found distributed in the cytoplasm and in all its prolongations, except the one designated the axis-cylinder process or the neuron: thus this process even at its base exhibits a character different from that of the other prolongations. By some methods of preparation the granules in the cytoplasm appear to be arranged in rows; these rows are sometimes concentric with the nucleus, but near the periphery of the cell they often extend into the prolongations, and if a significance is attached to their arrangement on the assumption that they indicate the pathways for the impulses, it is interesting to note that they occasionally enter the cytoplasm by one prolongation of the cell, and leave it by another, without interruption and without running close to the centre of the cell. When the cells of a single locality are treated with staining reagents, it is usually found that while some are deeply coloured others are not.

Age certainly is a factor in determining this colour reaction. The youngest cells, the neuroblasts, stain differently from those more developed; these latter differ according to their maturity, the condition of rest or fatigue, and also according to their physiological relations; those under control of the higher centres staining in a way different from those which belong to the reflex system.[1] All the prolongations arise primarily from the cell-body. Of these outgrowths there are two sorts: one, the dendrons, which branch in a tree-like manner into the surrounding substance, and the other the neurons, which have an even calibre and smooth contour, and from which the branches arise at right angles. Most cells have one neuron, some have two or even more, but the distinctions between these groups will be made later on. At its distal termination the neuron divides, in some cases many times, and according to circumstances the stretch of fibre between the cell-body and this terminal subdivision may vary from the fraction of a millimetre to half the length of a man; moreover, in many cases the neuron gives off small branches near its origin (Fig. 20, *A*), and these branches are apparently of much physiological importance. Besides the type of cell which forms the basis of the above description, Fig. 19 shows other forms (*B* and *C*), and the question arises as

[1] Kaiser, *Die Funktionen der Ganglienzellen des Halsmarkes*, Haag, 1891.

to how far they are to be regarded as variations of a single type. The forms represented by *A*, *E*, and *D* have but a single neuron, that is, are mononeuric, whereas the forms *B* and *C* are dineuric, having two neurons. The branching of the neuron near its origin

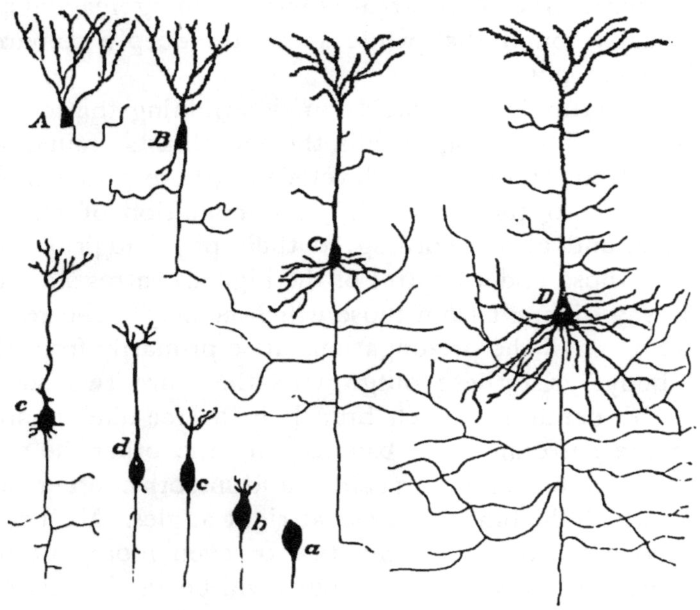

FIG. 21.—*A–D*, showing the phylogenetic development of mature nerve cells in a series of vertebrates: *a–e*, the ontogenetic development of growing cells in a typical mammal. In both cases only pyramidal cells from the cerebrum are shown. *A*, frog; *B*, lizard; *C*, rat; *D*, man; *a*, neuroblast without dendrons; *b*, commencing dendrons; *c*, dendrons further developed; *d*, first appearance of collateral branches; *e*, further development of collaterals and dendrons. (From S. Ramón y Cajal.)

is, in this connection, a suggestive feature. The above figure (21) shows the manner in which the branchings near the origin of the neuron occur in mononeuric cells. As can be seen, all the branches increase in number with the advancing position of the animal in the zoological scale, as well as with the advancing development of the individual cell.

Although these relations are easily recognised, it is not yet clear in how far all the branches of the neuron are alike; whether, for instance, each acquires the sheath of myeline which characterises the medullated nerve fibres, though it has been shown that at least some of these branches do become thus medullated (Flechsig).

The current conception of the nervous system demands an arrangement for the following physiological processes. In some way impulses must reach the cell-

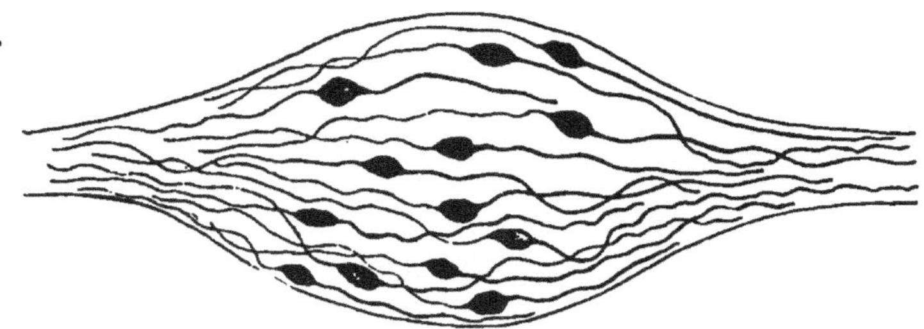

FIG. 22.—Spinal ganglion of an embryo duck, × 30 diameters. Composed of dineuric nerve cells. (Van Gehuchten.)

bodies within the central system. These bodies are masses of irritable cytoplasm, from which the discharge of energy may be much greater than that represented by the exciting stimulus. This discharge must be carried off by some pathway other than that by which the stimulus arrived. The simplest arrangement by which this is accomplished is the one found in Fig. 22.

This shows a spinal ganglion, containing typical bipolar or dineuric cells. In the usual disposition of such cells, one neuron extends to the periphery, and there terminates in such a way as to be exposed to some form of external stimulus. The stimulus acting on this distal end arouses an impulse which passes towards the cell-body, through it, and out by way of the other neuron, which in the case of these cells extends, as the dorsal nerve root, into the substance of the spinal cord, this root being the pathway by which the incoming impulses reach the central system. As a consequence of stimulation the cells themselves are changed, and it is therefore to be expected that the impulse leaving the cell would be different from that which entered it. Returning to the matter of form, the question arises whether the mononeuric cells, like those in the cortex (Fig. 21), are similar in fundamental structure to the more primitive dineuric cells, and have been derived from them. In favour of the view that they are thus similar, and derived, a few facts may be adduced. These ganglia in lower vertebrates are made up of dineuric cells, like those in Fig. 22. This form is assumed by the ganglion cells of mammals during their early development. In these latter, however, it is not permanent, for as growth continues the points of entrance of the neurons into the cell-body approximate until they come together, so that the cell-body finally stands at one side of the passing neuron. The change does not, however, stop here. From the neuron the cell becomes still further separated in such a way that it appears to stand at the end of a single stem, a stem which by its development is composed of those portions of the two neurons which were closest to the cell-body, and by virtue of its origin this stem should therefore contain two pathways.

The transformation can be beautifully seen in the developing Gasserian spinal ganglion of the guinea-pig. Here the cells when first formed are dineuric, as in the fish, and only gradually become modified to the extreme forms shown in the figure. It has been pointed out that the neurons in this instance do not always have a similar size, but differ so far that one can be regarded as the main stem of which the other is a branch. The important idea involved in this conception is, that the portion of the neuron appearing single and passing between the cell-body and the point of branching, contains by virtue of its origin two pathways for impulses, one going to the cell-body and the other coming from it. If this be correct, it is but a step to the cells whose neurons have

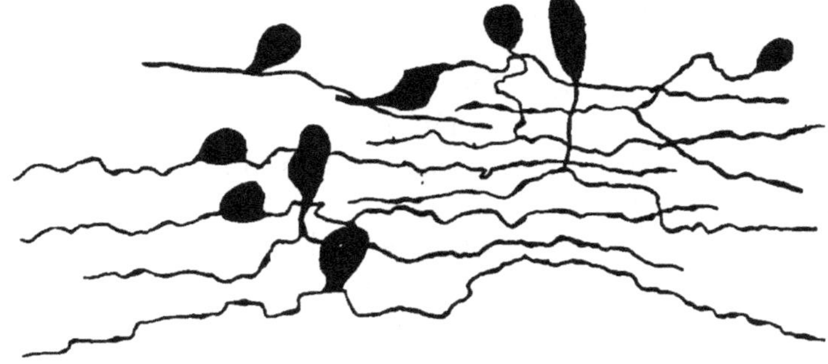

FIG. 23.—Dineuric changing into mononeuric cells. From the Gasserian ganglion of a developing guinea-pig. (Van Gehuchten.)

many branches, yet it must be admitted that but few facts are available in support of this view, involving as it does the notion that in the main stem the continuations of the branches are related to one another like isolated wires in a cable. The current explanation of the arrangement of the cell has been presented by Cajal and Kölliker. It is based on the notion that all the outgrowths of the cell are primitively similar in function. Accordingly, both the dendrons and the neurons are to be considered pathways for impulses, and the chief difference between them physiologically is assumed to depend on the fact that while the dendrons convey impulses towards the cell-body, the neuron carries them away from it.

An attempt to combine these two views, one based on embryology, and the other on the histology of the adult cell, would lead us here too far afield ; suffice it to say that both views assume that the incoming and outgoing impulses travel by separate pathways. Thus the suggestion that a given branch may contain more than one pathway is supplementary and not contradictory to the accepted view.

Though complete observations are wanting, we know in a general way that the nerve elements begin as small neuroblasts and grow large, the change in size involving all parts of the cell. The increase is at first rapid, then slow, but the time at which it ceases in any given cell cannot be exactly determined. It is, however, not possible to identify the growing period of all the cells with the growing periods of the individual, because the histological study of the nerve tissues shows that all the cells do not enlarge simultaneously.

There is reason to think that the nerve cells which begin to develop early increase in size as long as general growth continues, itself a period of uncertain length, while those cells which begin to develop later, increase up to the end of the same period. But the dates at which development may commence can be far apart, and hence the periods during which the cells are growing can have very different lengths. The discussion of the increase which occurs during development should be carried on in the light of all the changes which are then taking place. On examining a section of the developing spinal cord in which the germinal cells are still dividing, it will at once be seen that the germinal cells appear larger than the neuroblasts by which they are surrounded, and they also have a different character. The nucleus of the germinal cell or its immediate descendant is surrounded by cytoplasm clearly evident on all sides. In the typical neuroblasts the nucleus is large, and the cytoplasm seen only as an accumulation at one pole, being drawn out into a slender thread which represents the beginnings of the neuron. It has been shown by His, for example, that the mass of the neuroblast was, for a time after its formation, approximately equal to the mass of the parent germinal cell; the difference being that in the germinal cell the cytoplasm surrounded the nucleus on all sides, whereas in the neuroblast it was extended to form an embryonic fibre.

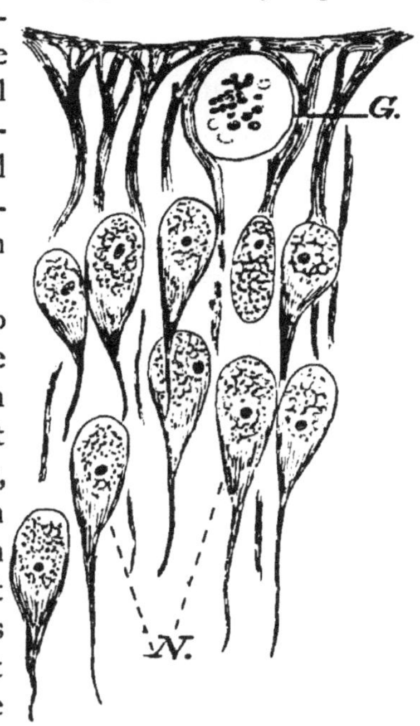

FIG. 24.—Portion of developing medullary tube seen in frontal section. Human, × 1,100 diameters (His). *G*, germinal cell; *N*, neuroblasts.

This relation enables us to take a germinal cell as the unit of volume from which to calculate the subsequent changes in size. Moreover, like the term cell, the term neuroblast in this discussion will be used to mean not only the nucleus with its surrounding cytoplasm, but also that portion of the cytoplasm which forms the young neuron. As the neuroblast develops, a striking change is caused in the cell-body by the apparently great increase in the cytoplasm, but this is really a small matter as compared with the increase in substance represented by the elongation and enlargement of the neuron. In this latter change the increase in the length of the neuron is the more important than the increase in diameter. It is known, for example, that continuous fibres pass from the cortex of the cerebrum to the lumbar enlargement of the spinal cord, and it therefore follows that certain neurons must extend over this distance, which, within the central nervous system, is the longest distance that is thus traversed. In the peripheral system the fibres stretch from their point of origin in the cord to the most distant portions of the limbs, and in man this is a greater distance than any within the central system. In each instance, be it remembered, the fibre reaching for this distance is the continuous outgrowth from a single cell-body.

Confining attention to the cell-bodies only, it is found that the volume of the germinal cell-body in the cord of the fœtus as compared with that of the nerve cell-body in the cord of the adult, may be as 1 to 100. From this cell-body proceeds the neuron, entirely wanting in all germinal cells, but which later on may, in some cases, acquire a length of three feet. Quantitatively this outgrowth is the most important portion of the cell; for even when the axis-cylinder alone is considered it may have a volume of more than 500 times that of the adult cell-body. There is thus developed a relation between the cell-body and its outgrowth which is quite unique, since the outgrowth is very many times more bulky than the body which gives origin to it. It will be necessary to use numerical statements in the subsequent argument, and it will therefore be desirable to give a record of the manner in which they have been obtained. The unit of measure employed by histologists is the micron. Its value is 0·001 mm., and its symbol the Greek letter μ. So far as possible, the following measurements will be given in this unit, since the figures will thus be most easily compared. In the accompanying table the value of the linear, square, and cubic millimeter is given in microns, and conversely that of the micron in fractions of a millimeter; also the weight of the milligramme in micro-milligrammes.

TABLE 32.—SHOWING THE RELATIONS OF THE MICRON TO THE MILLIMETER WHEN APPLIED TO THE MEASUREMENT OF LENGTH, AREA, AND VOLUME. ALSO THE RELATION BETWEEN THE MILLIGRAMME AND MICROMILLIGRAMME.

Length	1 mm.	=	1,000 μ
Area	1 sq. mm.	=	1,000,000 sq. μ
Volume	1 cu. mm.	=	1,000,000,000 cu. μ.
Length	1 μ	=	·001 mm.
Area	1 sq. μ	=	·000001 sq. mm.
Volume	1 cu. μ	=	·000000001 cu. mm.
Weight	1 mgrm.	=	1,000,000,000 μgrm.

This table will serve, among other purposes, as a reminder of the difficulty of conceiving either very small measures or very large numbers; so unacquainted are we with these extremes that most of us have no feeling for the correctness of the facts expressed in such terms, to say nothing of the difficulty in handling them.

The growth of the nerve cells may be briefly illustrated by a few measurements. His has determined the diameter of the germinal cells in man to be 11 μ, and the volume 697 μ (embryo R-length 5·5 mm., age 3–3·5 weeks).[1] The larger cells in the adult spinal cord have a diameter of 50 μ, and a volume of 65,312 cu μ. The ratio of these two volumes is a trifle less than 1–100, but for convenience we may take the round numbers. This liberty is justified by the fact that the very largest cells attain a diameter of more than 100 μ, and hence the figures for adult size given above are well within the limits. The larger nerve fibres are 10 μ in diameter, again a conservative figure, which gives 790 μ^2 as the area of the cross section, a relation which re-

[1] His, *Arch. f. Anat. und Physiol.*, 1889.

quires the length of the fibre having a volume equal to that of the cell-body 50 μ in diameter to be 828 μ. The long fibres passing between the cervical and lumbar enlargements are some 300,000 μ in length, so that a fibre of this size and extent would have 375 times the volume of the cell-body just mentioned, the combined volume of cell-body and neuron being ·00245 mm.^{3}, and the total weight, supposing it to have a specific gravity of 1·035, would be but ·0025 mgrms.

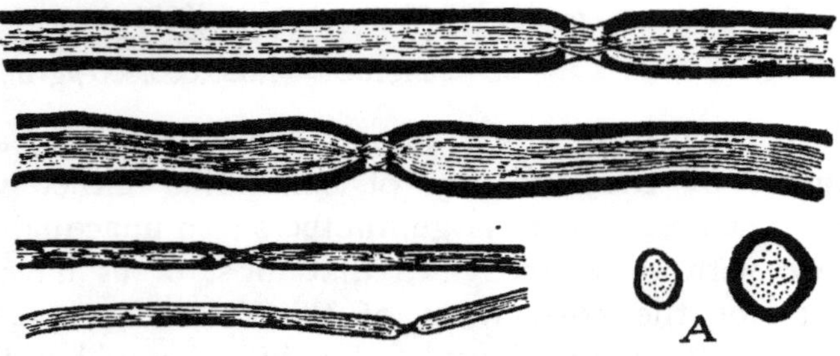

FIG. 25.—Longitudinal and transverse (A) sections of nerve fibres. The heavy border represents the medullary sheath, which becomes thicker in the larger fibres. Sciatic nerve. Human, × 400 diameters. (Modified from Van Gehuchten.)

The measurements show that in mature fibres of average size the area of the cross section is divided nearly equally between the axis-cylinder and medullary sheath, so that in the example given the axis-cylinder alone would have a volume 187 times that of the cell-body.

Similar calculations applied to the peripheral system of man show in even more marked manner the great proportional volume of the neuron, and in the lower mammals also analogous relations are found. At the same time it must be remembered that there are variations in both directions from the relations just described.

In the cells controlling the most distal segments of a limb, the disproportion between cell-body and neuron is even greater, since the cell-body is smaller and the neuron longer than in the example chosen. On the other hand, there are within the central system cells the neurons of which are so short that they probably have a volume even less than that of the cell-body.

So far as the matter has been studied, it appears that cells of small diameter have neurons of small diameter, while cells of large diameter generally have neurons of large diameter, but often the relations are by no means proportional. The two parts of the cell therefore vary in the same sense, but the variations are not strictly interdependent. Having determined that the volume of the nerve element does increase with age, it is desirable to also determine the manner in which this occurs. Kaiser's study of the cervical enlargement of the spinal cord illustrates this in part, although the measurements were made on the largest cell-bodies only.[1]

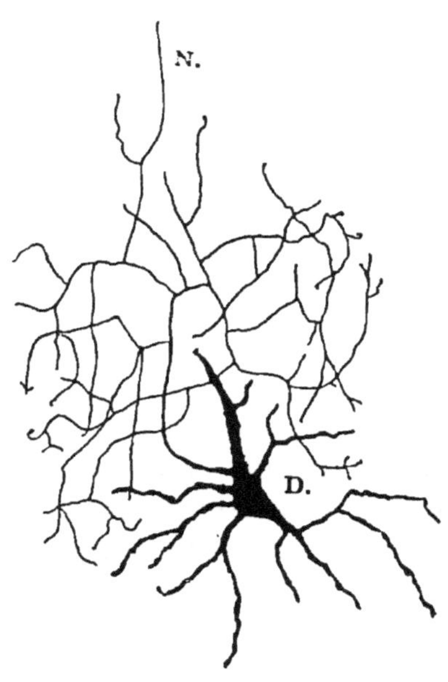

FIG. 26.—A cell with a short neuron giving off many branches. In such a cell the neuron is less in volume than the cell-body. This is the extreme form of the "central cell." (Ramón y Cajal.) D, dendrons ; N, neuron.

[1] Kaiser, *Die Funktionen der Ganglienzellen des Halsmarkes*, Haag, 1891.

TABLE 33.—SHOWING THE VOLUMES OF THE LARGEST CELL-BODIES IN THE VENTRAL HORN OF THE CERVICAL CORD OF MAN. (*Based on Kaiser's records of the mean diameters.*)

The volume $700\mu^3$, in the fœtus of four weeks, is taken from His, and the figures represent multiples of that volume.

SUBJECT.	AGE.	PROPORTIONAL VOLUME OF THE CELL-BODIES, $1 = 700\mu^3$.	TIME INTERVAL.
Fœtus	4 weeks	1	
"	20 "	17	
"	24 "	31	36 weeks
"	28 "	67	
"	36 "	81	
Child at birth ...	—	124	
Boy at 15 years ...	—	124	15 years
Man, adult	—	160	15 "

It shows a rapid increase in the volume of the largest cells during the earlier months before birth. Between birth and maturity the change is slight, and as a consequence these figures suggest that growth after birth is mainly due to some cause other than the further enlargement of the cell-bodies most developed at that time.

As a rule, when the size of nerve cells is stated in the text-books the measurements apply to the cell-body only, and the changes in this are expressed in the foregoing table, yet it is desirable to insist on the genetic connection between the cell-body and the neuron, and so the increase of the latter should be considered at the same time. In the mixed peripheral nerve of a new-born child the diameter of the fibres, including the sheath, was found (Westphal) to be 1·2–2 μ for the smallest, up to 7–8 μ for the largest, with an average of 3–4 μ. In the adult the large fibres are from 10–15 μ in diameter. The enlargement in diameter which is thus indicated is dependent on an increase in both the medullary sheath and the axis cylinder. The history of events is as follows: When first formed the axis cylinder is small in diameter, but soon increases. While this is transpiring the medullary sheath is formed, and grows rapidly in thickness, after which both constituents grow at such a rate that their respective areas, as seen in a cross section of the fibre, remain nearly equal to one another. The area of the cross section of any mass of fibres may therefore increase in two ways: either by the addition of new fibres which grow into it, or by an increase in the diameter of those already there present. This second method is shown in Table 34—based on the observations of Birge on the second spinal nerve of frogs—where the enlargement of the trunk is due to the increase in the size of the fibres, the increase in their number not being progressive in these instances.[1]

TABLE 34.—SHOWING THE INCREASE IN THE DIAMETER OF THE NERVE FIBRES COMPOSING THE SECOND SPINAL NERVE OF FROGS, WEIGHING 23 AND 63 GRAMMES RESPECTIVELY, AS COMPARED WITH ONE WEIGHING 1·5 GRAMMES. (*Birge.*)

WEIGHT OF FROG IN GRAMMES.	AREA OF CROSS SECTION OF 2ND SPINAL NERVE.	NUMBER OF FIBRES.	AVERAGE AREA OF FIBRES.	AVERAGE DIAMETER OF FIBRES.
1·5	·046 sq. mm.	986	46·6 sq. μ	7·6 μ
23·0	·105 ,,	1098	95·6 ,,	11·0 μ
63·0	·125 ,,	975	128 ,,	12·6 μ

[1] Birge, *Arch. f. Anat. u. Physiol.*, 1882. For an explanation of the failure of these nerves to increase in number of fibres as well as in diameter, the original paper should be consulted.

In this latter process the medullary sheath plays an important *rôle*. That there are axis-cylinders which do not have this sheath is probable, but its presence in connection with so many neurons makes it an important structural feature, and whatever view may be taken of its origin does not alter the fact that it is the companion of the axis-cylinder in all parts of the central system. The larger number of dendrons possessed by the more specialised cells suggests that in these cases there has been an increase in the number of channels through which the cell-body can receive stimuli, and by consequence its reactions become subject to more numerous modifications.

Thus far in this description the cell-body and the neuron have alone been described, while the dendrons have been neglected. However these latter are interpreted, whether as pathways or afferent impulses, or as merely nutritive channels, their importance is great, and variations in their number and size must be of profound significance, despite the fact that their total volume is very small. As will be seen by referring to the figures taken from Cajal (Fig. 21), the dendrons develop later than the neuron, and have also a different morphological character. They tend therefore to be more abundant in the older cells of the same individual, and in those taken from the animals high in the zoological scale.

In addition to the size of the nerve elements, it is also desirable to determine their number, either in the entire nervous system or in definite portions of it. If we knew the volume of the nervous system at any age, and at the same time the average volume of the nerve elements, the determination of number would be a simple matter of arithmetic. Unfortunately much uncertainty attaches to the data which must at present be employed, and the results are correspondingly unsatisfactory. Tabulated below are some figures which will be of assistance.

TABLE 35.—GIVING ESTIMATES OF THE VOLUME OF THE CENTRAL NERVOUS SYSTEM—ENCEPHALON AND SPINAL CORD AT DIFFERENT AGES.

Three-quarters of this volume is assumed to represent the nerve elements proper. For the first two records I am indebted to Professor F. P. Mall. The third is estimated.

SUBJECT.	AGE.	WEIGHT.	VOLUME OF NERVOUS SYSTEM—ENCEPHALON AND CORD.	
		Grms.	Vol., cu. cm.	¾ of this in cu. cm.
Fœtus ...	2 weeks	—	0·04	0·03
" ...	4 "	—	0·2	0·15
" ...	12 "	—	3·0	2·25
Child ...	At birth	[1] 381+4 385	376	282
Man ...	Adult	[2] 1360+26 1386	1340	1005

In the foregoing table the number of cu. cm. in the central system at different ages have been given, and then three-fourths of this taken as the volume of the nerve tissues proper, assuming that the other fourth is accounted for by the supporting and nutritive structures.

Meynert computed that there were 1,200 millions of ganglion cells in the cortex of the hemispheres, and that there were some ten millions of large cells in the cortex of the cerebellum. Adding to them the cells in the basal ganglia, the small cells everywhere, together with all those in the spinal cord, 3,000 millions would be a moderate estimate of the total number in the central system. Taking 3,000 millions as a working figure, the volume of the average cell (body and prolongations) in the central systems of an adult man is

[1] Specific gravity, 1·025. [2] Specific gravity, 1·035.

determined as follows: According to the table above given, this system contains 1,005 cu. cm. of nerve substance. The average volume of a cell would be then one 3000 millionth of this, which is ·00033 cu. mm., and taking the volume of the larger cells in the spinal cord as ·00246 cu. mm., then ·00033 cu. mm. would be about one-seventh of this figure. If, relying on histological evidence, it is assumed that the production of cells ceases at the end of the third month of fœtal life, and that the volume of the nervous system at that time is about 2·25 cu. cm., the volume of the neuroblasts would be 750 cu. μ. It was found previously that the germinal cells had a volume of 687 cu. μ, or nearly the same figure; the neuroblast soon becoming larger than the cell from which it was derived. The importance of this calculation is the following: Assuming that there are 3,000 millions of nerve elements in the central system of 1,005 cu. cm., and that these are condensed into a nervous system whose volume is only 2·25 cu. cm., the size of each of the elements is approximately the size of the germinal cell, as determined by His from direct observation, and the enlargement which each of these 3,000 millions of elements requires to undergo in order to attain the average size found in the adult system amounts to but one-seventh of the volume which is possessed by the larger cells of the adult spinal cord. In order that the nervous system, with a volume of 2·25 cu. cm., increase to 1,005 cu. cm., the number of elements remaining the same, all the individual elements in the central system must increase in average volume (1,005 ÷ 2·25) 447 times. The largest elements in the nervous system do increase 10,000 times in volume, hence the average increase here demanded is not large in comparison with that which may occur. Since many neuroblasts increase far beyond the average volume just indicated, there is, in order to account for the adult system, not only no need of arrangement for the production of new cells after the third month of fœtal life, but it must also be assumed that a large number of the elements already formed undergo but a very slight development, otherwise the final mass of the central system would be far more voluminous than it is. The relation of the mass of this system at the 12th week of

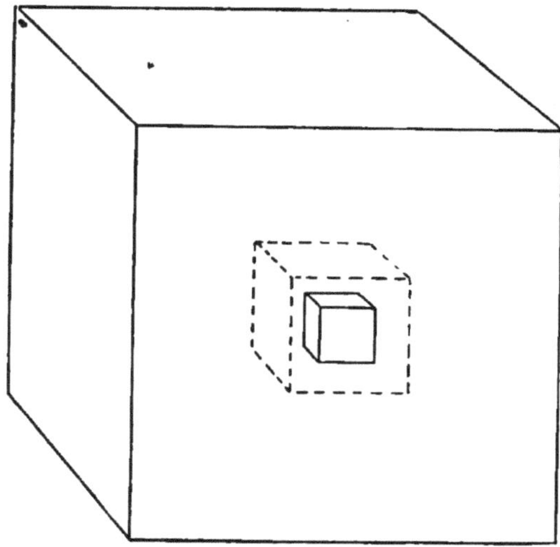

FIG. 27.—Cubes illustrating the relative volumes of the central nervous system at the 12th week of fœtal life and at maturity. If the largest cube represent the volume of the central system of an adult, the smallest will represent that of a fœtus at the 12th week, both reduced to one-tenth of their actual volume. The cube indicated by the dotted line represents the estimated volume of the central system in the fœtus of the 12th week.

fœtal life and at maturity is illustrated in Fig. 27, which shows in solid lines two cubes having one-tenth their true volume, and a third, indicated by the dotted lines, which has the actual volume of the central system as estimated for a fœtus of the 12th week.

The determination of the number of neuroblasts occurs so early in the history of the individual, and under such uniform conditions, that it is very difficult to regard the environment as possessed of much power to cause variation in this respect, and for this reason among members of the same race a high degree of constancy in this character is to be anticipated. The influence of the surrounding conditions becomes much more effective during the later stages of development that accompany the enlargements of the elements already formed, and it is during this period that adaptive modifications may occur.

CHAPTER VIII.

DEVELOPMENT OF NERVE ELEMENTS.

Time of development—The elements at birth—Schiller's observations—The frog—Observations by Birge—Origin of sensory fibres—The frog compared with the mammal—Significance of medullary substance—Regeneration — Means by which the brain-weight changes with age—Influence of stature on brain-weight—Significance of brain-weight in different races—In the two sexes—Among animals—Anatomical relations alone explained.

THE constructive development of the nerve elements consists in an increase in size, in the formation of the abundant cytoplasm and outgrowths, and in chemical modifications which affect differently the several subdivisions of the cell. The cells are first clearly characterised in the spinal cord, then in the medulla oblongata, and finally in the cerebral hemispheres and cerebellum. But this modification begins in different groups and in different members of the same group at different times. Kaiser, in his valuable observations on the arrangement of the nerve cells in the cervical enlargement of the mammalian cord has made some interesting studies on exactly this point.[1] His table (36) shows the number of developed nerve cells in the similar portions of the cervical spinal cord of man at different ages, beginning with the fœtus at the fifth month.

[1] Kaiser, *Die Funktionen der Ganglienzellen des Halsmarkes*. Haag, 1891.

TABLE 36.—SHOWING THE NUMBER OF DEVELOPED CELLS IN THE CERVICAL ENLARGEMENT OF MAN AT DIFFERENT AGES. (*Kaiser.*)

AGE.	NUMBER OF NERVE CELLS.
Fœtus—16 weeks.	50.500
„ 32 „	118,330
New-born child.	104,270
Boy. 15 years.	211,800
Male. Adult.	221,200

Thus he is able to show that in this locality the developed cells increase in number from the fifth month of fœtal life to maturity, the number doubling between the fifth and ninth months of fœtal life, and again between birth and the completion of growth. The first period of increase probably centres about the sixth month of fœtal life, during which time the absolute increase in the central system is large, and the second period in all probability occurs during the first seven years of childhood, since by the end of that time the central system has attained nearly its full weight. The difference in the number of cells observed in the fœtus of thirty-two weeks and in the new-born child is probably not a mere error of observation, but indicates a more rapid development of the nerve cells in the former case, yet at the same time it has little significance for the absolute number which will ultimately be developed, because even the higher figure is only a trifle more than one-half that representing the total number of nerve cells in the adult. To determine the number of cells usually functional at maturity, the observations on grown persons must be many times repeated. There is also no direct indication of the age at which the development of the nerve cells in man ceases, yet if the changes in the cortex of the adult, which I shall later describe, are rightly interpreted, growth in that locality must be going on between the twentieth and fortieth years, and, judging from Venn's observations on the enlargement of the head, is extended over the first half of that interval at least.

Much interest has always attached to the determination of the time when the central system was structurally completed. As might be anticipated, the various portions are completed at very different periods, according to their functions, and hence all statements on this head require to be made in detail. Schiller[1] counted the number of nerve fibres in the oculo-motor nerves of cats at birth and at later ages. His results are given in summary in the table which follows.

TABLE 37.—SHOWING THE AVERAGE NUMBER OF FIBRES FOUND IN EACH OF THE OCULO-MOTOR NERVES OF CATS, FROM BIRTH TO EIGHTEEN MONTHS OF AGE. (*Schiller.*)

AGE OF SPECIMEN.	MEAN NUMBER OF FIBRES.	EXTREME VARIATION IN NUMBER OF FIBRES.
New-born. A. B. C. (Average of 3 cases)	2942	2905–2980
1 month. D. E. (Average of 2 cases) ...	2961	2946–2976
4 months. F.	3007	2995–3016
12 „ G. (Mother of A. B. F.)	3013	3002–3019
18 „ H.	3035	3020–3050

This table shows that between birth and maturity there is in the cat a very slight increase in the number of fibres composing the oculo-motor nerve. The author explains the slightly smaller number found at the earlier ages as due to the small size of the youngest fibres, which are thereby made exceedingly

[1] Schiller, *Compt. Rend. Acad. d. Sc.*, Paris, 1889.

difficult of enumeration. The early numerical completeness of this nerve is remarkable, and the explanation of this may possibly depend on the fact that it is a nerve with fixed physiological functions, the demands upon it being similar at all ages. It is at the same time conceivable that nerves whose functions become more complicated with age might be, even in the same animal, far less complete at birth. It is possible, too, that in man the condition of the oculo-motor nerve is the same as that found in the cat, yet it has just been shown that in man the cervical enlargement of the spinal cord is only half developed at this time.

Still pursuing the questions of number, there are a series of interesting facts, which have been determined by Birge in the case of the frog.[1] He sought an answer for two questions: first, does the number of nerve fibres in the ventral roots of the spinal cord of the frog increase with the size (age) of the individual? and, second, does the number of developed fibres in the ventral roots correspond with the number of developed cells in the ventral horns of the cord, thus presenting the relation to be expected if each neuron were the outgrowth of a single cell-body? Both these questions were answered in the affirmative. The examination was made upon seven frogs, ranging in weight from one and a half to one hundred and eleven grammes. The total number of nerve fibres in the ventral roots of the smallest frog, weighing 1·5 grammes, was 5,984, or 3,990 fibres for each gramme of frog. In the heaviest frog, weighing 111 grammes, the total number of fibres in the ventral roots was 11,468, or 103 fibres for each gramme of frog. It is plain from this that on the whole as the frog grows heavier its muscles and skin have a proportionally smaller number of fibres controlling them. Taking the

[1] Birge, *Archiv. f. Anat. u. Physiol.*, 1882.

five frogs which were intermediate between these extremes, it is found that from the smallest frog up to the heaviest each additional gramme of frog is accompanied by an absolute increase of about 50 nerve fibres in the ventral roots. The following table exhibits the results in greater detail.

TABLE 38.—SHOWING THE NUMBER OF FIBRES IN THE VENTRAL ROOTS OF FROGS OF THE SAME SPECIES, RANGED ACCORDING TO THEIR INCREASE IN BODY-WEIGHT. (*Birge.*)

FROG. WEIGHT.	NO. OF FIBRES.	EXCESS OF WEIGHT.	EXCESS FIBRES.	NO. OF FIBRES PER GRAMME EXCESS OF WEIGHT.
Grammes.		Grammes.		
1·5	5984	—	—	—
9·5	6481	8	497	62
23	7048	21·5	1064	50
63	8566	61·5	2582	42
67	9492	65·5	3508	53
87	10004	85·5	4020	47
111	11468	109·5	5484	51

From these observations it is concluded that the rate at which the nerve fibres are produced in a frog before it reaches 1·5 grammes in weight, is many times that at which the production takes place in the larger frogs. It is certainly remarkable that after this earliest stage the rate of increase is quite constant. An examination of the figures in detail shows that the increase in the number of fibres is greatest and most persistent in the cervical group (Roots I., II., III.), next in the lumbar group (Roots VIII., IX., X.), while in both cases the number of fibres in the largest frog is more than double that in the smallest, whereas in the thoracic region (Roots IV., V., VI., VII.) the corresponding ratio is only 1–1·3. To guard against a too hasty interpretation of these results, it is to be remembered that a nerve grows not only by the increase in the number of its fibres, but also by the enlargement in diameter of the fibres at any time present, and therefore the observations on number should be supplemented by measurement of the area of the nerves at different ages.

It was further shown that the fully-developed large nerve cells in the ventral horns of the frog's cord, each corresponded to a single nerve fibre in the ventral roots, and hence the results support the view that there is a one-to-one relationship between them. As the number of ventral fibres increases, the number of developed cells in the cord also increases, and there is thus determined for the entire cord of the frog growth relations similar to those which Kaiser found in the cervical enlargement of man. Some attention was also given to the dorsal roots. Birge counted the fibres in these in the case of two frogs, as shown in the Table 39.

TABLE 39.—SHOWING THE NUMBER OF FIBRES IN THE DORSAL ROOTS OF TWO FROGS HAVING DIFFERENT WEIGHTS. (*Birge.*)

WEIGHT.		NO. OF DORSAL ROOT FIBRES.
23 grammes	......	7562
63 ,,	......	10670

The frogs are identical with the specimens of 23 and 63 grammes, the number of fibres in whose ventral roots has been given in Table 38. In both cases the figures show the dorsal roots to have the greater number of fibres, and the rate of increase is 77 fibres for each gramme of frog—a rate which is 50 per cent. greater than the average rate for the ventral roots.

There is good reason to think that the fibres in the dorsal roots, with perhaps a trifling exception, take their origin from the cells of the spinal ganglia. These cells when developed are typically dineuric, and hence the number of fibres emerging from either end of the ganglion should be equal to one another, and also equal to the number of cells in the ganglion. Nelson[1] first, and then Hodge, undertook to determine the number of fibres in the dorsal roots, and also the number of cells in the associated ganglia. Nelson found in small frogs the number of cells more than twice that of the fibres. Hodge has obtained figures which are similar.[2] These figures suggest that all the cells which are there present have not yet sent out their neurons, and that in the larger and older animals the proportion of the developed cells in the ganglion would increase, and hence the proportion of cells to fibres would diminish; but for this explanation good evidence is still wanting.

By comparison it appears that during early life the development of nerve cells goes on more actively in man than in the frog, but the difference is one of degree only, and, as we shall later see, in the cerebral cortex at least it continues in man for a very long time. The data to be presented will be useful in interpreting the size of the brain, but as a preface to such an interpretation the relative development of the grey and white substance is first to be determined.

During early fœtal life there is no distinction between white and grey matter, the entire system being grey, but with increasing age the medullary substance appears, the change first occurring in the cord. According to the estimates of De Regibus, the grey matter forms 58·5 per cent. of the entire weight of the cerebral hemispheres. What is designated "grey" matter in these cases is really a mixture of cells and fibres, and the cells,

[1] His records were destroyed when the Science Hall of the University of Wisconsin was burned in 1884.

[2] Hodge, *Am. Journ. of Psychology*, 1889.

excluding their neurons, do not certainly form more than one-tenth of the entire weight of a sample of grey substance. The cellular portion weighs therefore 6 per cent., and the white portion, consisting of the neurons, weighs 94 per cent. of the entire encephalon. Employing the previous calculations whereby the mass of the medullary substance is about one-half of the mass of the neuron, then one-half of this weight, or 47 per cent., is medullary substance proper. At maturity, therefore, this substance alone has a weight nearly twice that of the entire encephalon at birth. But no special argument is needed to show the importance of this factor in the determination of brain-weight. Its quantity is probably variable within wide limits, yet the degree of this variation has never been investigated.

That form of forced growth which is called regeneration, the capacity for which is so early lost by mammals, is very remarkable, and will repay attention as a property of nerve cells. A distinction should be made, however, between the regeneration which involves the reproduction of new specialised elements by cell division, and the regeneration of a lost portion of a cell by that part which still persists. When in a mammal, for example, a peripheral nerve is cut, it can regenerate, and this regeneration can involve the production of a large quantity of material and also be successively repeated as many as three times. It is a remarkable function of the peripheral nerves, and contrasts sharply with the conditions within the central system, where such regeneration does not occur. So far, then, as the central nervous system is concerned, it cannot renew itself in the sense of regenerating elements, either lost or injured, whereas injured peripheral nerves can be repaired.

From the facts concerning the number and size of the structural elements, it is evident that the enlargement of the nervous system is due in the first instance mainly to an increase in the number of neuroblasts formed, and in the second, to the size which they later attain, and that to the increase in size are due those growth changes which are usually recorded. The non-nervous tissues in the central system are assumed to represent 25 per cent. of its total volume, and to have the same proportion at different ages, this being but a gross approximation, allowable only in the absence of better data. The interpretation now to be attempted is based on the notion that in organic evolution structure precedes function in the sense that the slightly specialised elements are not at once utilised, and that a partial development of nerve cells may result in their presence in considerable masses, without implying a corresponding physiological activity. This being granted, the further inquiry is as follows:—

1. *By what means does the brain of the new-born attain the weight found in the adult and decrease again during old age?* The weight of the male child's brain at birth being taken as 372 grammes, and that of the adult male as 1,360 grammes, there is an absolute increase of 988 grammes to be explained. The brain increases in weight during this time by the increase of its non-nervous constituents and by the enlargement of its nerve elements. These latter changes are dependent on the increase in the mass of the cell-body and cell outgrowths, especially the neuron, and in the acquisition by the neuron of a medullary sheath. Calculation shows that the absolute mass of the medullary substance is the chief source of increase in weight during this period. The weight increase in the nerve elements proper is due to the enlargement of those cells which at birth are small, the addition of new cells being excluded since their formation ceased before birth. In answer to the first question, it can therefore be stated that between birth

and maturity the increase in size of the brain is due in part to the enlargement of the cell elements, but in greater measure to the formation of the medullary substance accompanying this enlargement. The loss of weight in extreme age probably depends on destructive processes affecting the medullary substance, accompanied by a general decrease in the proportion of water.

2. *Why do tall persons have heavier brains?* Individuals who are above the average height are, as a rule, also above the average in weight, and it would help greatly in this inquiry if the histological basis for differences in weight and stature were known—whether, namely, they were for the most part due to an increase in the number or the size of the cells. The cranial cavity and spinal canal would tend to be enlarged in either case, and the size of these cavities is one condition affecting the size of the enclosed nerve structures. Also it seems probable that the variations in number or size would affect both the supporting and the nerve tissues alike, and hence that the larger central system and the large cavities enclosing them are the result of similar variations in both—namely, an increase in the *size* of the structural elements.

3. *What significance is to be attached to the fact that the brain-weight is different in different races?* When individuals belonging to different races are thus compared, it appears probable that differences in the number of the cell elements may be of importance, but there is no direct evidence on this point, and since in different races variations in brain-weight follow in general variations in the body-weight, it is necessary therefore to attribute the differences to the same conditions which are important in the case of the persons of different stature, remembering, on the one hand, that racial differences are more marked among adults than among infants; and, on the other, that variations in the absolute

number of cell elements must be determined even before birth, at a time when in other respects the individuals are most similar. The provisional picture, therefore, to be formed of the brains belonging to those races least capable mentally is that of one in which the number of cell elements is approximately similar to that in the most capable races; but many of these elements being but partially developed, the organisation of the brain is less perfect, though the size is not thereby greatly reduced.

4. *What significance is to be attached to the difference in brain-weight existing between men and women?* Keith[1] has shown that among monkeys the males have the heavier brains. The fact that among mammals generally the males are the larger and heavier, renders it probable that this relation of the brain-weight will be commonly found in this class. Whatever inference is drawn, therefore, from this relation will have an application far beyond the limits of the human species. The encephalon in the two sexes has a nearly similar specific gravity, and the development of its subdivisions is in almost the same proportion. In every case, however, where average weights have been compared, the male encephalon has been found to be heavier than that of the female. At birth the encephalic weight is nearly the same for the two sexes, and the greater part of the differences found between them at maturity is developed during the first seven years of life. In macrocephalic, large, medium, small, and microcephalic brains the weight relation between the sexes is similar, though both the absolute and relative differences increase with increasing weight, just as in the body-growth of the two sexes. Yet the microcephalic brains are most instructive in this connection. A minimal amount of com-

[1] Keith, *Journ. of Anat. and Physiol.*, vol. xxix., 1895.

plexity in the encephalon is assumed to be necessary for the maintenance of life. This minimum is presumptively represented by the least developed group of microcephalics, yet even here the female brains are below those of the males in weight, although it must be assumed that in both the structural complexity is probably alike. If this is so, the difference in weight between the two groups must depend on the fact that the structural elements in the encephalon of the female are smaller than those in the male. What is true of this group is probably true of the others, and the size of the nerve elements rather than variations in their number is therefore to be regarded as the principal factor in determining the difference in brain-weight. Size has a double significance here. In the first place, the larger cell has the larger proportion of unstable cytoplasm, more stored energy. In the second, and this is a condition peculiar to the nervous system, increase in size accompanies increase in organisation ; but these points will be taken up later.

5. *What value is to be given to the size of the brain in different groups of animals ?* As has been shown, small dogs have smaller brains than large dogs, but in proportion to the size of the entire animal, the smaller dogs have the larger brains. The same is true of any other similar group of animals. Moreover, the animals which are absolutely the largest are in general the more intelligent. The explanation appears to depend on the fact that the mass of the individual elements is not proportional to the mass of the body, so that if, for example, the cell elements from a middle-sized dog be taken as a standard, and those from the extreme forms be compared with it, those from the smallest dogs will be found proportionately large, while in the largest they are proportionately small. Second, it depends on the fact that

those reactions of the central system which form the basis for the general intelligence of the species require a given complexity in this system, which complexity in turn demands a large number of elements. No matter, therefore, how small the size of the entire animal, its nervous system must possess this minimal number of cells, and in the smallest representatives these are of a relatively large size. For this reason the smaller members of such a group have proportionately the largest central systems, while the largest members have numerically the most complex central system, and therefore the anatomical basis for a better intelligence.

In general, then, the variations in the weight of the adult human brain just mentioned are best explained by regarding them as mainly dependent on the size of the constituent nerve elements. The nervous system of animals with a smaller brain-weight than man probably contains both fewer cells, and of these the developed cells are also fewer in number. What the relation is in those with a brain-weight above that of man is not known, but the probabilities are in favour of the numerical superiority of man here also. Before concluding, however, I desire to emphasise two points: first, that the explanations just suggested apply to anatomical differences only, to the mass of encephalon, its physiological capabilities being another matter; and second, that the expression of these differences in the terms of the constituent cells has the advantage of preventing a too hasty inference from the brute figures indicating weight differences to the subtle physiological possibilities of this complex organ.

CHAPTER IX.

ARCHITECTURE OF THE CENTRAL NERVOUS SYSTEM AT MATURITY.

Architecture at maturity—Shape of skull—Shape of brain—Effect of deformation—On weight—Relative development of lobes—Causes of change in shape—Bilateral symmetry—Weight of right and left hemispheres—Decussation—Segmentation—Areas of spinal cord—Cephalic development—Brain and cord in animals.

A DESCRIPTION of the form and construction of the central system, as found in the adult, must precede the study of the important changes by which they have been attained. Since the days when interest in such matters was first aroused, the shape of the head has received its share of attention, and various efforts have been made to interpret it. The usual varieties among normal skulls are designated as brachycephalic, or broad, and dolicocephalic, or long. As a matter of experiment it appears that the brachycephalic skulls are more capacious, and we should expect this, perhaps, since they are most nearly spherical in shape.

In the dolicocephalic heads the lengthening does not increase the volume of the cranial cavity to a degree sufficient to compensate for the loss consequent upon the lateral compression. These differences in the capacities of the two sorts of skulls are slight, and have thus far failed of any general interpretation. Aside, however, from the differences in the shape of the skull, due to intrinsic causes, there exists a large group of artificially deformed skulls.

By the application of pressure to the head of the growing child many races seek to alter the shape of the skull. The results are often striking, as can be seen by consulting the accompanying Figs. 28 and 29, which show some of the extreme modifications that have been observed as a consequence of pressure applied in various ways.

These peculiar skulls are of interest because the brain conforms itself to the shape of the skull, and therefore in such cases the brains deviate from the usual form as much as the skulls which enclose them.

A paper recently published by Ambialet[1] on the deformation of the heads of people living about Toulouse gives perhaps the most consecutive account of such deformation, and the effect of it on the brain. In Fig. 28, 2, is shown a skull deformed in the way which Ambialet describes. On the ground that the head of the child should be protected from cold, it was formerly a custom in this part of France to wind it with a compressing bandage in such a way as to produce the unusual lengthening which the figure shows. The shape of the brain is accordingly altered, but there is no indication that the number or relative development of the gyri is thus rendered peculiar. Provided the child is able to adjust itself to the first stages of development, its intelligence is not noticeably diminished. At the same time the brain is thus forced to grow into an abnormal

[1] Ambialet, *La Déformation Artificielle de la Tête dans la Région Toulousaine*, Toulouse, 1893.

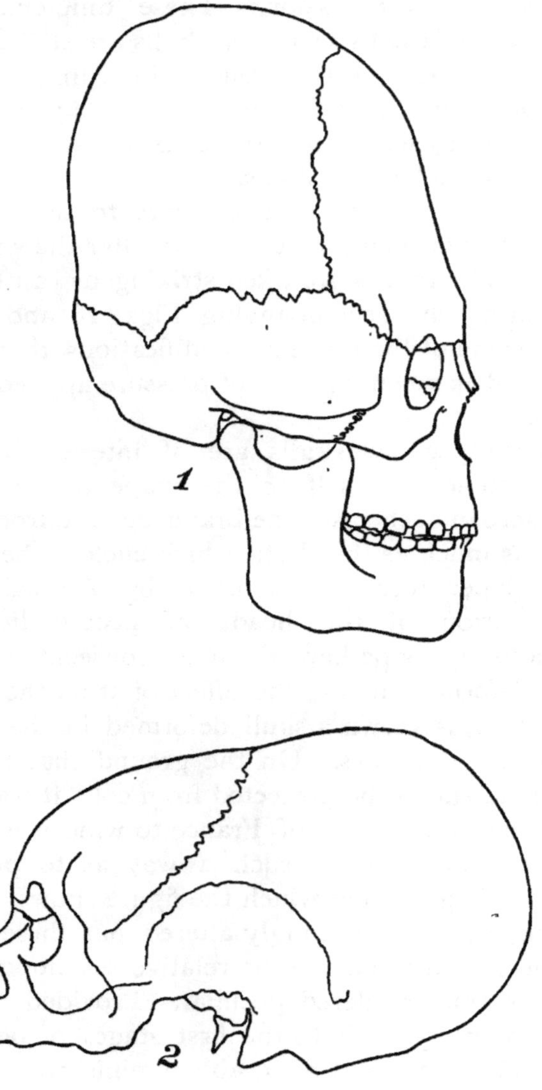

FIG. 28.—Deformed skulls. 1, cuneiform deformation, Natchez Indian. (Topinard, after Gosse.) 2, deformation from Toulouse, Department of Haut-Garonne, France. (Ambialet.)

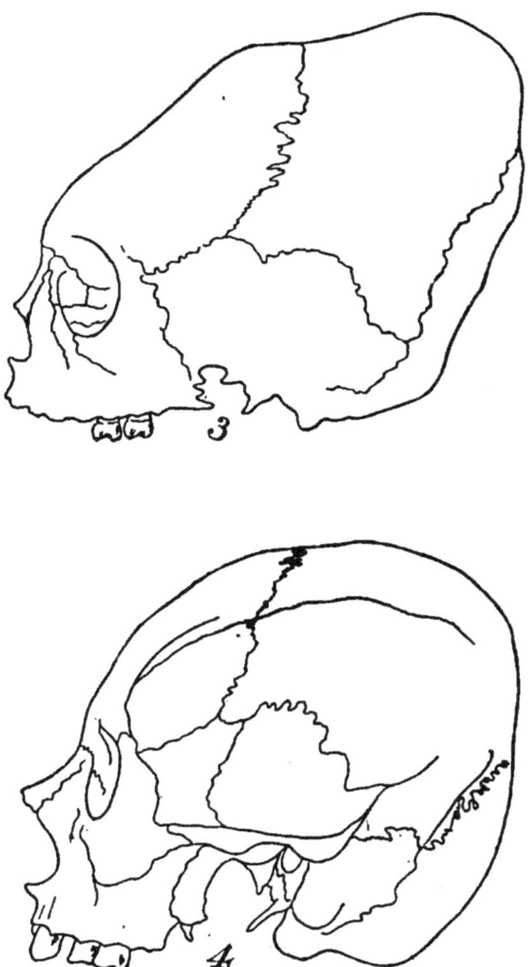

FIG. 29.—Deformed skulls. 3, deformation from the Island of Malikolo, Polynesia. (Topinard.) 4, deformation in a Maronite, Syria. (Topinard.)

form, and it is not at all surprising to find that compensation in growth has not been complete, and that the brains of these persons are slightly under weight. Table 40, from Ambialet, shows the weight of the encephalon and its parts for males, Table 41, the same for females. The weights given in Tables 40, 41, and 42 were all obtained by the method of Broca. According to this method, the brains are deprived of the pia, and allowed to drain for five minutes before being weighed.

TABLE 40.—GIVING THE BRAIN-WEIGHT IN THE CASE OF MALES AFFECTED BY THE DEFORMATION OF THE SKULL PRACTISED ABOUT TOULOUSE. THE CEREBELLUM AND STEM HAVE BEEN WEIGHED TOGETHER. (*Compiled from Ambialet.*)

The average weight of the normal encephalon (French males between 65–85 years—Broca) is given below for comparison.

MALES.

NO. OF OBSERVATION	AGE.	DEFORMITY.	WEIGHT.		
			Encephalon.	Hemispheres.	Cerebellum and Stem.
I.	72 years	Exaggerated	1091 grms.	940 grms.	146 grms.
II.	62 „	Moderate	1260 „	1073 „	180 „
III.	70 „	„	1095 „	945 „	145 „
IV.	77 „	„	1295 „	1122 „	165 „
V.	70 „	„	1370 „	1210 „	150 „
VI.	81 „	Marked	1210 „	1019 „	175 „

I.–VI. Average weight ... ... 1220 grms.
Normal males, 65–85 years (Broca), average weight, 1266 „

TABLE 41.—GIVING THE BRAIN-WEIGHT IN THE CASE OF FEMALES AFFECTED BY THE DEFORMATION OF THE SKULL PRACTISED ABOUT TOULOUSE. THE CEREBELLUM AND THE STEM HAVE BEEN WEIGHED TOGETHER. (*Compiled from Ambialet.*)

The average weight of the normal encephalon (French females between 65–85 years—Broca) is given below for comparison.

FEMALES.					
NO. OF OBSER-VATION	AGE.	DEFORMITY.	WEIGHT.		
			Encephalon.	Hemispheres.	Cerebellum and Stem.
VIII.	68 years	Marked	1080 grms.	920 grms.	151 grms.
XI.	65 „	Moderate	1129 „	977 „	140 „
XII.	66 „	„	1158 „	1011 „	142 „
XIII.	72 „	„	1112 „	955 „	147 „
XIV.	64 „	Exaggerated	838 „	760 „	127 „
XV.	78 „	Extreme	1190 „	993 „	194 „

VIII.–XV. Average weight ... ... 1093 grms.
Normal females, 65–85 years (Broca), average weight, 1110 „

The average weight of the encephalon is given in each table, and for comparison also the normal weights in both sexes between sixty-five and eighty-five years of age. These later observations were made by Broca on normal French subjects, and hence are the most directly comparable with the figures of Ambialet. It is seen that the "moderately" deformed male brains (average weight 1,255 grms.) are slightly below Broca's figure; whereas the corresponding brains among the females (average weight 1,133 grms.) are slightly above. It is therefore to be concluded that this deformity when moderate is associated with compensatory growth, but when the deformity becomes excessive this compensation is usually incomplete.

It is very interesting to find, however, that if the hemispheres (mantle plus the basal ganglia) are divided into their lobes in the manner in which Broca was accustomed to separate them, the average proportional weight of these lobes is similar to that found in the individuals taken as a standard; for the slight deviations which are here noted are not of any significance.

The mode of division is shown in Fig. 30.

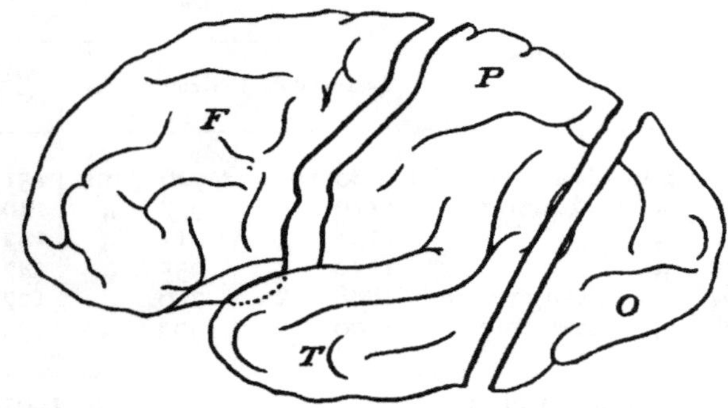

FIG. 30.—Left hemisphere of a cerebrum divided into lobes according to the method of Broca. *F*, frontal lobe; *O*, occipital lobe; *P*, *T*, parieto-temporal lobe.

TABLE 42.—GIVING THE PROPORTIONAL WEIGHT OF THE LOBES (FRONTAL, PARIETAL, OCCIPITAL) OF THE DEFORMED CEREBRUM, DIVIDED ACCORDING TO THE METHOD OF BROCA, AND COMPARED WITH THE PERCENTAGES FOUND BY BROCA IN NORMAL FRENCH SUBJECTS BETWEEN 70–90 YEARS OF AGE. (*Ambialet.*)

No. of Cases.	Frontal lobes.	Parietal lobes.	Occipital lobes.
MALES.			
6	43·5	47·3	9·1
Normal average (Broca)	42·9	45·8	11·2
FEMALES.			
6	42·8	46·7	10·17
Normal average (Broca)	43·7	46·2	10·1

Table 42 gives the records as taken from Ambialet's paper, and below in each case similar averages derived from the study of normal brains observed by Broca. (French subjects, 70–90 years of age.)

Changes in the shape of the brain are not, as might at first sight appear, the result of direct pressure, for, under normal conditions, the encephalon is nowhere in immediate contact with the skull, but the influences by which the brain is reduced below the normal size must be those of disturbed nutrition. Even the form of the brain is to some extent independent of pressure. For instance, the gyri develop on the surface of a growing brain, even though the skull has been removed, and when one hemisphere has been taken out of the cranium the other extends to only a slight degree across the middle line. It appears, therefore, that in all these instances it is not the mechanical conditions which are most effective.

After the general shape, the bilateral symmetry of the central system is the most important feature. The conception is well founded that the plane which divides the body into two symmetrical portions divides the nervous system in the same way, but in neither case is the symmetry perfect. Granting such symmetry to be the typical condition, what is to be said in the case of persons with skulls which are manifestly very far from symmetrically developed? In part such deviation shown by the outside of the skull may fail to be represented in the cranial cavity, but in general asymmetry without means asymmetry within, and it is to be expected that in such an individual one half of the encephalon would be heavier than the other, and that the influence of this inequality would extend to all the parts with which it was connected. Excluding the cases which have just been mentioned, it may be fairly asked concerning the individuals whose encephalic development appears to be symmetrical, what degree of perfection is thus accomplished. The results of examination with this idea in view may be given in two ways: either by summing the weights of all the right and left hemispheres in a given series, and from these finding the average amount by which one hemisphere exceeds the other; or by determining the number of cases in which the two hemispheres weigh the same, and also the number in which one half exhibits an excess of weight.

TABLE 43.—TO SHOW THE AVERAGE AMOUNT IN GRAMMES BY WHICH EITHER HEMISPHERE EXCEEDS THE OTHER.

RIGHT HEMISPHERE.			LEFT HEMISPHERE.		
Observer.	Subjects.	Excess grms.	Observer.	Subjects.	Excess grms.
Broca... ...	Adult males	1·93	Boyd	Both sexes, all ages.	3·7
,, ...	,, females	0·03			
Danielbekof	Children— Boys, 1 month	0·72			
,,	Girls ,,	0·65			

Taking the simplest method, we have an expression of facts like this in the above table (43), where, for example, all the male brains weighed by Broca had on the average right hemispheres which were 1·93 grammes heavier than the left; whereas Boyd found the left hemispheres to show an average excess of 3·7 grammes. Taking the weight of the male brain as 1,360 grammes, then 3·7 grammes is only ·27 per cent., which is really very small. Using the second arrangement, the facts are expressed by the following table (44) from Franceschi.

TABLE 44.—GIVING THE NUMBER OF CASES IN WHICH THE HEMISPHERES WERE EQUAL IN WEIGHT, OR ONE OF THEM IN EXCESS, IN A SERIES OF ITALIAN BRAINS WEIGHED BY FRANCESCHI.[1] WEIGHTS WITHIN ONE GRAMME OR LESS ARE CALLED EQUAL.

	AGE.	NO. OF CASES.	NO. OF CASES IN WHICH THE WEIGHT OF		
			Left Hemisphere > the Right.	Right Hemisphere > the Left.	Left Hemisphere = the Right.
Males...	10–87 years	157	49	51	57
Females	10–87 „	144	43	46	55

This table does not indicate any tendency of one hemisphere to be constantly the heavier. So long as the differences are as small as those here noted, and so long as different observers obtain results directly opposed to one another, as the first table shows, the figures do not furnish a good basis for induction. That the results of different observers are opposed is perhaps to be explained by certain technical difficulties. When it becomes a question of passing a knife through a brain so as to divide it into two halves, there is found, in a complicated form, the same difficulty that presents itself on trying to divide a straight line into equal parts. Experiments show that in the case of a right-handed person attempting this division of the cerebrum by drawing the knife from heel to point, the edge being down, the tendency is to leave the larger portion of the brain to the right-hand side of the blade. Thus it would happen that if a perfectly symmetrical brain were laid upon its ventral surface, with the frontal end away from the operator, and the attempt made to

[1] Franceschi, *Bull. d. Sc. Med. di Bologna*, 1888.

exactly divide it, the right hemisphere would probably be the heavier. If the occipital end were away from the operator, the reverse result would follow. Until, therefore, it is reported exactly what method was employed in making the division of the brains, it will not be safe to attach any importance to the recorded differences in the weight of the two hemispheres.

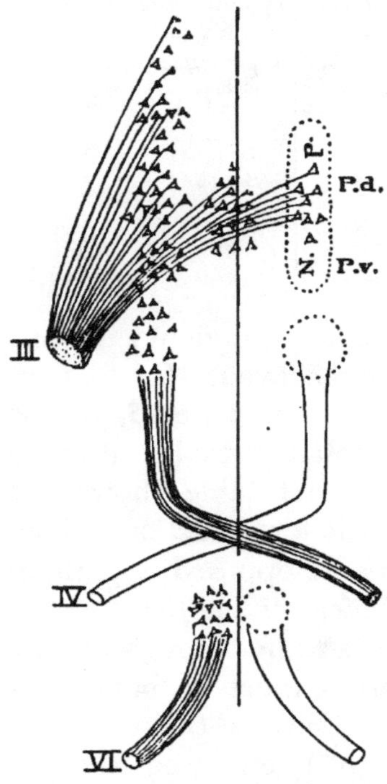

FIG. 31.—Illustrating the partial and complete decussation of the fibres of the third and fourth cranial nerves, and the absence of decussation in the case of the sixth. III., root of the third cranial nerve; IV., of the fourth; VI., of the sixth.

Asymmetrical development may in certain cases be the normal. In one instance, at least—viz., the cerebellar cortex of the cat—an asymmetrical development is a regular occurrence, and preliminary observations show the same to be true for the sheep. In this division of the brain the molecular layer of the cortex shows an average thickness in the right hemisphere, which is always less than that in the left.[1] Perhaps this difference is associated with the twisting of the vermis to the right side; and as this asymmetrical position of the vermis is by no means confined to the cat and sheep, but appears in other vertebrates, we may expect to find similar differences existing in them also.

[1] Krohn, *Journ. o Nervous and Mental Disease*, 1892.

One general constructional feature in the mammalian system is the crossing of the middle line by the neurons of many cells. This arrangement brings the cells located on one side of a median vertical plane into connection with the structures located on the other side. This capacity for decussation is confined to the elements of the medullary system, whereas the outgrowths of the ganglionic system remain on the same side of the body as that on which they originate. The fact that these decussations are completed rather late in the growing period suggests why they should be highly variable, and thus at times contribute to asymmetry in form. Within the central system the decussations of the optic fibres and the pyramidal tracts are the most familiar examples. The different cranial nerves exhibit this arrangement to varying degrees, and while the decussation is complete in the case of the fourth nerve and absent in the case of the sixth, it is partial in the third. (Fig. 31.)

Outside of the central system, decussations are rare, but have been described for the peripheral nerves going to the electric organ in the torpedo (Fritsch) and in the superior laryngeal nerve of the tortoise (Weir Mitchell). As illustrating the variability in these decussations, Flechsig's study of the pyramidal tracts in man may be cited.[1] It is known that at the crossing of the pyramids certain groups of fibres pass the middle line, and ultimately enter the lateral column on the opposite side of the spinal cord. Others do not decussate at this point, but pass into the ventral columns, ultimately crossing at lower levels. Flechsig found that practically the two extreme cases conceivable might be present, viz., that all the fibres might pass into the lateral column or

[1] Flechsig, *Leitungsbahnen im Gehirn und Rückenmark*, Leipzig, 1876.

nearly all of them into the ventral columns. Finally the mass of pyramidal fibres in the two halves of the cord is in many cases very unequal. No better example of variability in construction can be given, and from this arrangement it is evident that if there is an original bilateral symmetry of the central system it can be modified by the subsequent growth processes.

Besides the division of the nervous system into two symmetrical halves there is another manner of dividing it, which is of much morphological importance. Comparative anatomy suggests that it is formed by a series of segments or equivalent portions arranged in linear series. In some lower vertebrates this character is quite apparent, and opposite to the places at which the spinal nerves are attached the cord is somewhat enlarged owing to the larger number of fibres which enter the cord at these points, and also to the increase in the cell-bodies, both in the number and in the size.

In various ways this primitive arrangement is modified. Among higher vertebrates the principal sense organs located exclusively in the head assume a greater relative importance, and the reactions of the entire organism become more and more subject to them. This depends upon the fact that the various centres distributed through the spinal cord become connected with the cells lying at the head end in such a manner as to be somewhat controlled by them. These connections are mediated by bundles of fibres, which, traversing as they do the length of the cord, disturb the segmental arrangement. Moreover the great development of nerve elements in the cord at the regions where the nerves controlling the limbs are given off, causes a very considerable enlargement, extending through a number of primitive segments. As a result of all these modifications, the primitive segmental character of the medullary tube is much obscured in man.

The quantitative relations of grey and white matter composing the human cord are illustrated by the following diagram. The curves are intended to illustrate in terms of square millimeters the area of the cross-section of the cord at the level of each spinal nerve. In this figure the white substance and the grey are contrasted with one another as well as the united sectional areas of the spinal nerves. The curves show that, corresponding to the lumbar and cervical enlargements, a great increase

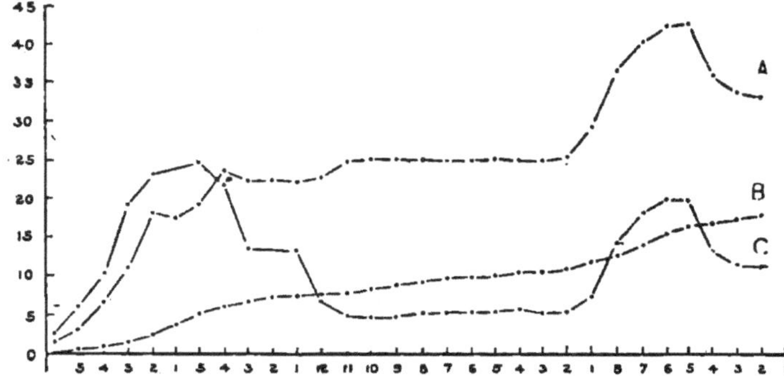

FIG. 32.—Diagram showing by curves the area of several portions of the cord. Areas indicated in square millimeters on the vertical axis. The level of the spinal root is marked on the horizontal axis, the first cervical being that recorded at the extreme right. A, the area of the total white matter at the different levels; B, the *united* area of the spinal nerve roots beginning at the sacral end of the cord; C, the area of the grey matter at the different levels. (Compiled from measurements by Stilling.)

in the area of the grey substance occurs, while the white substance, taken as a whole, increases in area from the caudal to the cephalic end, with a special expansion in the cervical region. The united sectional area of the nerves show an almost steady increase, which is slightly accentuated at the points where the nerves for the limbs join the cord.

In these curves no attempt has been made to represent the grey and white matter at different levels in the enlarged cephalic end, the area of which in the cross-section would of course be many times that of the spinal cord. There is perhaps no more striking feature in the architecture of the mammalian nervous system than precisely this development of the encephalic portion, as compared with either the cranial nerves or the spinal cord. Special interest attaches to this, since there is good reason to consider the better development of the cephalic end as an index of greater intelligence. Plain as this fact is in a general way, it is still difficult to express it concisely. For the present the data indicating the weight of the cord, as compared with the weight of the encephalon, are alone available. Though the weight of the human encephalon is superior to that of all but a very few animals, the human spinal cord is surpassed in weight by almost all the larger mammals. This relation, however, varies greatly with age. In man also there is found a change in the weight of the spinal cord, whereby the adult has proportionately a larger cord than the child at birth. This relation is expressed in Table 45. Table 46 shows the relation of the cord to the brain in a series of mammals at birth and maturity.

TABLE 45.—SHOWING THE WEIGHT OF THE SPINAL CORD AS COMPARED WITH THE ENCEPHALON IN MAN AT DIFFERENT AGES. (*Compiled from Mies.*)

AGE.	WEIGHT OF SPINAL CORD.	WEIGHT OF ENCEPHALON.	
		Males.	Females.
Fœtus—3 months	1	18	
" 5 " ...	1	101	
At Birth	1	116	113
In Adults	1	51	49

TABLE 46.—SHOWING THE CHANGE WITH AGE IN THE PROPORTIONAL WEIGHT OF THE SPINAL CORD AS COMPARED WITH THE ENCEPHALON IN A SERIES OF MAMMALS. (*Compiled from Mies.*)

ANIMAL.	PROPORTION AT BIRTH. ENCEPHALON-CORD.	PROPORTION IN ADULT. ENCEPHALON-CORD.
Rabbit ...	9–1	2–1
Cat	14–1	3·5–1
Dachshund ...	19·2–1	5–1

In several small birds, as the starling and siskin, the relation between the encephalon and the cord is as 10–1; while in some fishes and most of the amphibia the spinal cord is heavier than the brain.[1]

[1] Mies, *Neurolg. Centralbl.*, 1893.

CHAPTER X.

ARCHITECTURE (*CONTINUED*): PERIPHERAL NERVES—CENTRAL CELLS—CEREBRAL CORTEX.

Cranial nerves—Dorsal and ventral plates of the medullary tube—Fissuration—Measurements—Area of cortex—Thickness of cortex.

It has been pointed out by the older investigators that as compared with many animals, the cranial nerves in man are smâll, and that this small size is not only relative, but absolute. So far as these nerves are those of special sense, it might be thought that the diameter of the channel would be important in determining their physiological value. As a matter of fact, comparative anatomy does not support this notion, and physiology has long since suggested that it is the organisation of the elements which form the central termination of the afferent fibres that is of greatest importance. In this connection is given a chart showing the area in cross-section of all the sensory and motor nerves in man. (Fig. 33.)

Unfortunately there are no adequate observations on animals to compare with these, so that the relations between the areas of the cranial and spinal nerves cannot be contrasted with those in man.

There yet remains another way of dividing the central system according to certain features shown most clearly during its earlier development. This method of division may be roughly described as the result of passing a plane through the medullary tube in such a way as to

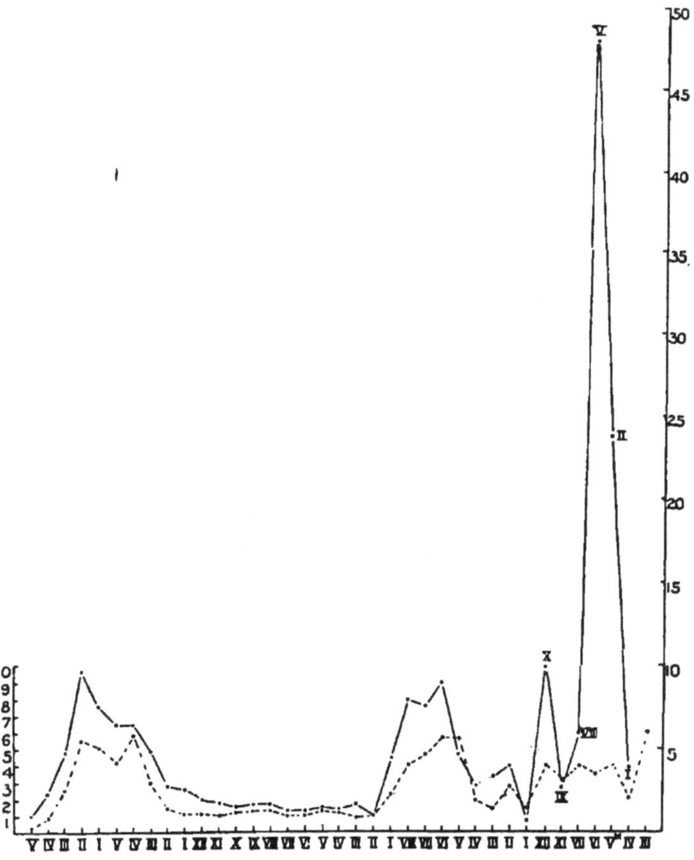

FIG. 33.—Diagram to show in square millimeters the areas of the sensory and motor nerves of the entire nervous system. The motor nerves are represented by the broken line, the sensory by the solid line. At the right the numerical designations for the sensory cranial nerves are put on the curve. The area of the sensory nerves is regularly greater than that of the motor, and proportionately increased towards the head end. (The curve for the spinal nerves is based on figures by Stilling.)

separate the ventral from the dorsal portion. The idea

of this separation and the significance attached to it have come mainly from the observations of His.[1] The following diagram will serve to illustrate the principal points.

This diagram represents a cross-section of the developing spinal cord. Within is the cavity forming the central canal. The surrounding walls exhibit an indentation on either side, which roughly divides each side into two portions, a dorsal and a ventral. The two dorsal portions are the dorsal plates (Flügelplatten-His) and the two ventral portions the ventral plates (Grundplatten-His) of this tube. The ventral plates contain all the nerve cells whose neurons grow out of the central nervous system. The dorsal plates contain no such nerve cells; they receive, however, the fibres which grow in from the cells forming the spinal ganglia, and in both plates are found the central cells whose neurons are confined within the limits of the medullary tube. The groove which separates these two important portions of the wall can be followed from one end of the nervous system to the other. On tracing this dividing groove through its entire length, it is found that the relations of the dorsal

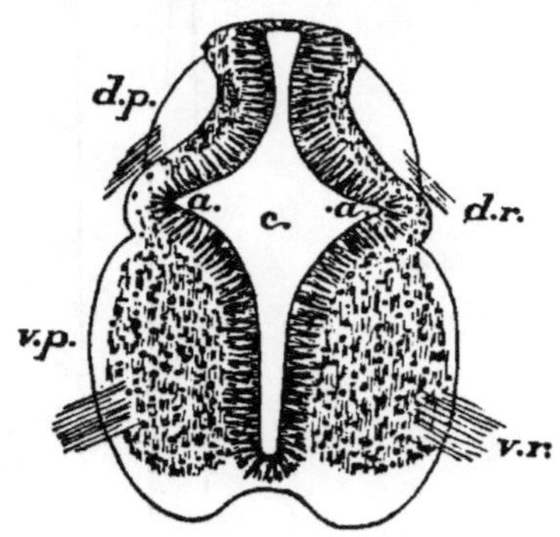

FIG. 34.—Cross-section in the cervical region of a fœtal human spinal cord, at the sixth week, × 50 diameters. (Kölliker.) *c.*, central canal; *a.a.*, groove separating the two plates; *d.p.*, dorsal plate; *v.p.*, ventral plate, in which alone are developed nerve cells producing neurons that leave the central system; *d.r.*, dorsal nerve root; *v.r.*, ventral root.

[1] His, *Abhandl. d. Math. Phys. Cl. d. Königl. Sächs. Gesellschaft der Wissenschaften*, 1889.

and ventral portions are exhibited by a diagram like the following,[1] in which the shaded portion represents the part that has been derived from the ventral plates, while the unshaded portion represents that derived from the dorsal plates.

Examining the encephalon as thus divided, we find that the cerebellum, the dorsal portions of the

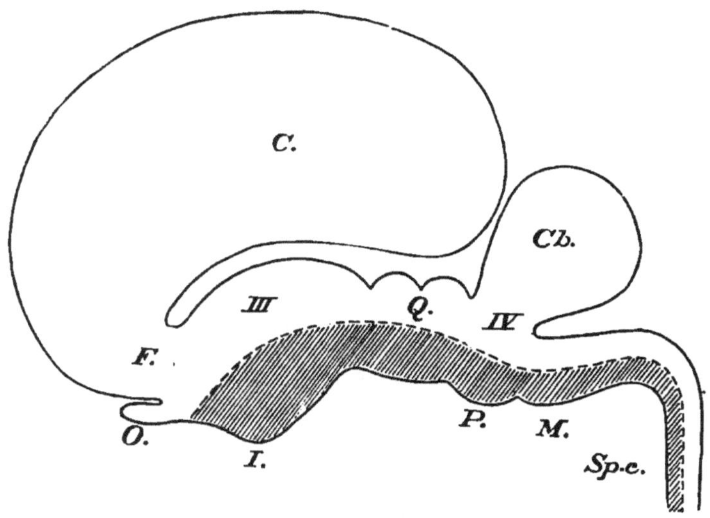

FIG. 35.—Schema showing the encephalon and cord. The unshaded portion is that derived from the dorsal plate, and the shaded that from the ventral. (From Minot.) *C.*, cerebrum; *Cb.*, cerebellum; *F.*, foramen of Monro; *I.*, infundibulum; *M.*, bulb; *O.*, olfactory lobe; *P.*, pons; *Q.*, quadrigemina; *Sp.c.*, spinal cord; *III.*, third ventricle; *IV.*, fourth ventricle.

quadrigemina and thalamic ganglia, as well as the entire mass of the cerebral hemispheres, are derived from the dorsal plates. These parts of the encephalon are therefore the homologues of the dorsal horns of the spinal cord. This portion of the cord is occupied by

[1] Minot, *Pop. Science Monthly*, 1893.

central cells, the function of which is the distribution of the incoming impulses within the central system. Accordingly, similar cells having a similar duty are to be expected in the homologous portions of the encephalon—a conclusion which is supported by the fact that on other grounds just this function has been attributed to them. When the nerves associated with these two plates are studied, their arrangement is found to be equally characteristic. Taking the central system segment by segment, the sensory nerves are more numerous and have a greater area than the motor. In man, as we pass cephalad, the superiority of the sensory nerves become most marked, and while the leg with its great mass of muscle and extent of skin is hardly better innervated than the arm, the innervation of the small muscle mass and skin area of the face is far more perfect than that of either. Stilling[1] has estimated the number of fibres in the ventral roots of the spinal cord as 300,000, in the dorsal roots as 500,000, or 1–1·6, thus indicating the great numerical excess of the latter. From the figures collected by Vierordt it appears that, if the olfactory tract and optic nerve be reckoned with cranial nerves, as they usually are, then the motor fibres among the cranial nerves stand as 86,000 against 2,548,000 sensory ones, or 1–30. When the innervation of the arm and that of the leg in man are compared similar relations are found. The leg weighs three times as much as the arm, and has twice the superficial area, yet the arm has nearly as great a nerve supply, and proportionately the superiority of the arm is greatest as regards the sensory nerves (Fig. 33). As the head is approached, therefore, there is not only a proportional and absolute increase in the number of the central cells occupying

[1] Stilling, *Neue Untersuchungen über den Bau des Rückenmarks*, 1859.

the region of the dorsal plate, but also an increase in both the motor and sensory innervation of the body, especially on the sensory side.

The encephalon is by far the most interesting portion of the nervous system, and therefore has been scrutinised with the greatest care, even the folds which mark the surface of the cerebrum having been studied in detail. These folds, or gyri, are separated from one another by the sulci or fissures. The questions which have been raised concerning this feature are, first, whether the arrangement of the folds is a constant one, and second, if it is not constant, whether any conclusion can be drawn from the variations which occur. At one time it was assumed that variations in the intelligence could be measured by this means, and that there was also a difference in these folds according to sex. At present the question is also discussed whether certain groups, criminals and the insane for instance, have brains differently marked from those of normal persons. Finally it has been thought that the study of the brains of various races might bring out important differences. The accompanying figures, taken from Eberstaller [1] and Schäfer,[2] represent the appearance of a Western European brain; that of Eberstaller is the best schema of the lateral aspect that has been published. By the aid of such a chart it has been practicable to test the questions to which allusion has just been made, but these tests have failed to establish any constant peculiarities of fissuration characteristic of sex, race, or social class.

Since, however, the hemispheres of the brain are, in the fœtus, smooth, and only gradually become marked by fissures, it is observed that early disturbances in their growth, which later are almost invariably associated with

[1] Eberstaller, *Das Stirnhirn*, Vienna, 1890.

[2] Schäfer, Quain's *Anatomy*, vol. iii. pt. i., 1893.

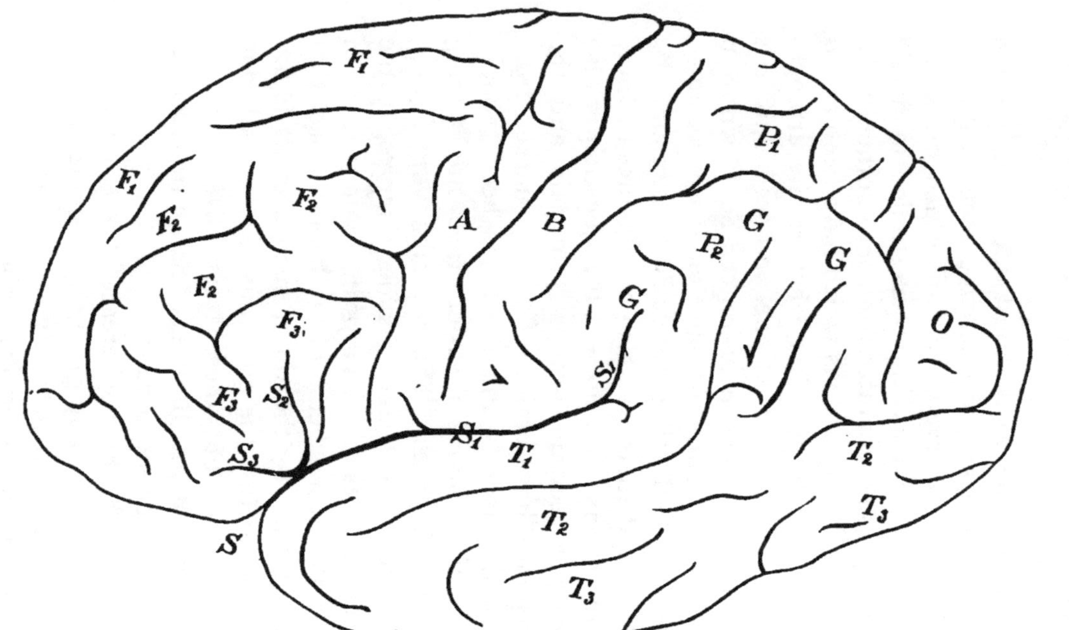

FIG. 36.—Lateral view of a human hemisphere showing fissures with intermediate gyri. With the exception of the Sylvian fissures the gyri alone are designated. (Eberstaller.) Gyri: *A*, anterior central; *B*, posterior central; *F*, superior frontal; F_2, middle frontal; F_3, inferior frontal; *O*, occipital lobe; P_1, superior parietal lobe; P_2, inferior parietal lobe; T_1, first temporal; T_2, second temporal; T_3, third temporal. Fissures: *S*, stem of the Sylvian fissure; S_1, Sylvian fissure; S_2, anterior ascending branch; S_3, anterior horizontal branch.

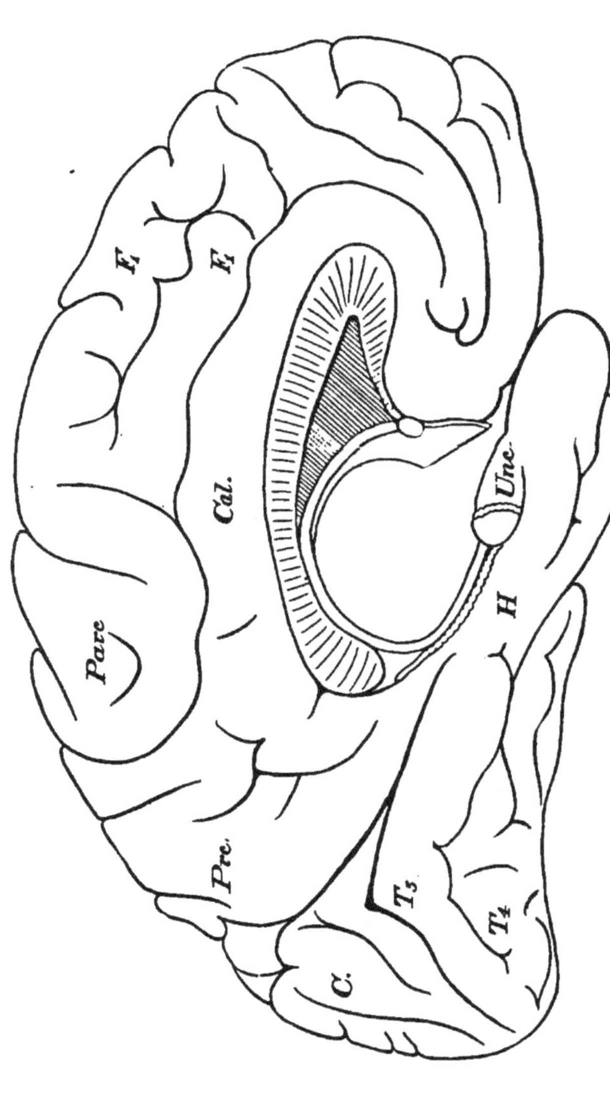

FIG. 37.—Mesal view of a human hemisphere, showing fissures with the intermediate gyri. (Quain's *Anatomy*.) *C*, Cuneus; *Cal.*, Callosal gyrus; F_1, Superior frontal; *H*, Hippocampal; *Parc.*, Paracentral lobe; *Prc.*, Precuneus; T_4, Fourth temporal; T_5, Fifth temporal; *Unc.*, Uncinate gyrus.

disturbances in the intelligence, are also accompanied by abnormalities in fissuration. But further than this it is hardly safe to go. On the hemispheres of the human cerebrum a number of statistical studies of fissuration and measurements have been made. These are mainly to be found in the works of Giacomini, Eberstaller, Cunningham, and Mingazzini.[1] As the result of comparing the hemispheres in the two sexes, or from the two sides of the same brain, there have been noted by various observers a number of peculiarities in fissuration. These features of the cerebral surface have a range and value remotely to be compared with that of the features of the face, but they are even less constant and less easy to interpret. There are no characters by which the sex of a given brain can be recognised with certainty; nevertheless investigators are generally agreed that the male brain tends to be more extensively fissured. But concerning such matters as the extent of the frontal lobe or the relations of the Sylvian fissure, and the peculiarities of other important fissures, the differences, if general, are too slight to make practicable their presentation here.

With the folding of the surface and the production of the gyri are connected several growth problems of great interest. Anatomically there are two conditions to be fulfilled in the development of the cerebral hemispheres. In the first place there must be provision for a large number of fully developed cells, and in the second place these cells must become physiologically connected with one another. It is plain that these two processes do not necessarily go hand in hand, and we may have every combination, from a large number of cortical cells adequately associated, to a small number incompletely associated. On the number and development of the

[1] Mingazzini, *Il Cervello*, &c., Torino, 1895.

cell-bodies the extent of the cerebral cortex is mainly dependent, while to the associating fibres is due the mass of underlying white substance. A brain possessed of an extensive cortex the elements of which are incompletely associated, can therefore be a much folded brain, because the layer of substance forming the cortex is so extensive that in order to have it applied to the surface of the cerebrum it must be thrown into many gyri. On the other hand, the development of the associating fibres, increasing as it does the central mass of the white substance, gives a larger surface to which the cortex may be applied, and in so far tends to diminish its folds.

The significance of fissuration as an index of intelligence receives no support from comparative anatomy, since the brains of ruminants are much more convoluted than those of the dog, while the heavier and more intelligent birds have brains that are nearly smooth. As has been suggested the fissuration of the brain surface depends upon several variables, and the problem must first be simplified by analysis before a general conclusion is attempted. Precisely at this point, failure to recognise the complexity of the anatomical conditions involved has led to some doubtful inferences. It has been asserted by those most interested in the study of the brains of criminals, that these brains were theromorphic in their surface markings, that is, showed a similarity in this respect to the lower animals, especially the carnivora. This would imply that such brains had been arrested at some early stage. But since the human brain in the course of its development does not pass through a carnivorous phase, it follows that any such peculiarities are fictitious.[1] It has

[1] Cunningham, *Contribution to the Surface Anatomy of the Cerebral Hemispheres*, Dublin, 1892.

been also asserted that in the criminal brain the fissures ran together in such a manner that the channels formed by them are more continuous than in the case of normal individuals. It requires still to be proved that this is characteristic of the brains of this class, for non-criminal brains differ very markedly in the degree to which the fissures are confluent. In view of what has been said just above, this greater or less confluence of the fissures may be interpreted in various ways, and while confluence, so far as it goes, indicates a large area of the cortex, it may have quite different meanings in large and small brains.

By various devices the extent of the folded surface ot the hemispheres has been measured. These measurements have been made either directly or by determining the volume of the cortex, and then, assuming that it had an average thickness, by making an estimate of its extent. Among the earlier observations were those by Wagner,[1] who undertook to determine the area of the cortex in the brains of Fuchs, the pathologist, and Gauss, the mathematician, and to compare this area with that found in the brains of a labouring man and a woman of ordinary intelligence. Unfortunately this determination was not made until after the brains had lost from about 27 to 40 per cent. in weight by their preservation in alcohol, and had consequently undergone some diminution in volume, so that the figures given for the areas in these several cases apply to the brains after this treatment. Some slight data are available by which a rough correction of these results can be made (Donaldson), and thus corrected,[2] the weight of the cortex appears in Table 47.

[1] H. Wagner, *Massbestimmungen aer Oberfläche des grossen Gehirns*, Inaug. Diss., Göttingen, 1864.

[2] Donaldson, *Journ. of Morphol.*, 1894

TABLE 47.—SHOWING THE RELATIONS BETWEEN THE FRESH WEIGHT OF THE HEMISPHERES AND THAT OF THE CORTEX. BASED ON THE DIRECT MEASUREMENTS OF H. WAGNER.

The average thickness of the cortex is taken to be 2·9 mm. The weights are therefore proportional to the area.

	WEIGHT OF HEMISPHERES.	WEIGHT OF CORTEX.	PERCENTAGE WEIGHT OF CORTEX.
Fuchs ...	1312	743	56·6
Gauss ...	1306	731	55·9
Artisan ...	1114	630	56·5
Woman ...	1037	667	64·3

The weight is a much more manageable measurement than the area, and has therefore been employed in the tables. In normal persons the average thickness of the cortex is not open to wide variations, and hence the weights are proportional to the area. With these observations are to be compared those made by Calori on Italian subjects.[1] His results are summarised in the following table (48), which indicates that the weight or extent of the cortex in the male is greater than in the female, and also that when the brachycephalic individuals are compared with the dolicocephalic, the former are found to possess the greater mass of cortex. Finally, under the direction of Giacomini, De Regibus determined the area of the cortex by the last method mentioned above. I have compiled from his figures the accompanying table (49), but used for their reduction constants slightly different from those which he employed, thus obtaining, I believe, more correct results, again expressed as weights.

[1] Calori, *Mem. della Accad. della Sc. dell' instituto di Bologna*, 1870.

TABLE 48 (SIMILAR TO TABLE 47).—BASED ON THE RECORDS OF CALORI.

The extent of the cortex was calculated geometrically. The records are separated according to sex and shape of head.—B. Brachycephalic. D. Dolicocephalic.

Cases.		Weight of Hemispheres.	Weight of Cortex.	Percentage Weight of Cortex.
Males.	19. B.	1168	731	62·5
	15. D.	1155	690	59·7
Females.	3. B.	1005	635	63·2
	3. D.	971	595	61·2

TABLE 49 (SIMILAR TO TABLE 47).—BASED ON THE RECORDS OF THE PERCENTAGE OF WATER BY DE REGIBUS. FOUR CASES. ITALIAN BRAINS.

Cerebrum.	Weight of Hemispheres.	Weight of Cortex.	Percentage Weight of Cortex.
1	1277	720	56·3
2	1194	661	55·3
3	1152	633	54·9
4	1067	562	52·6

So far as these figures show anything, it is that the larger brains have a larger area of cortex; and further, that as here given there is a very fair concordance among the results obtained by the different observers. When it can be stated in what manner the total brain-weight is to be interpreted, it will be possible also to say what significance is to be attached to the extent or weight of the cerebral cortex. In the meantime it is interesting to note that of the total area of the cortex, almost exactly one-third is found on the exposed surface of the hemispheres, while the other two-thirds form the walls of the fissures. It is evident from this that other things being equal, the increase in the total length of the fissures would indicate an increase in the cortical extension. The relation of the area which is exposed to that which is sunken is shown in Fig. 38.

FIG. 38.—Diagram illustrating the extent of the cerebral cortex. The outer square shows a surface one-twenty-fifth of 2,352 sq. cm. in extent. The inner square has two-thirds of this area, and is the proportion of the cortex sunken in the fissures. 2,352 sq. cm. is approximately the area of the entire cortex in a male brain weighing 1,360 grms.

The actual total area, say 2,352 cm., is something more than one-fifth of a square meter. The diagram is one-twenty-fifth of the actual area represented by 2,352 sq. cm. The inner square has two-thirds of the area of the outer one, so that the space between the outlines of the two squares represents one-third of the area of the larger square, or one-half the area of the smaller one, and is the proportion of the cortex exposed on the outer surface of the hemispheres. In the tables by Wagner the relative development of the cortex in the two hemispheres was noted, and it was found to be more extensive on the right side in the case of Fuchs and Gauss, whereas in the woman and labouring man that of the left side was greater. In dealing with the extent of the cortex by methods of direct measurement, its thickness was not determined, and the earlier investigators were ready to admit that an important datum was therefore lacking. Since that time, however, a number of elaborate investigations on the thickness of the cortex have been made. The method of determining the average thickness of the cortex is to measure samples of this layer from a large number of localities and then fuse the results. The accompanying table, giving the figures obtained by the principal investigators, shows that the different observers are by no means in accord.[1]

TABLE 50.—SHOWING THE AVERAGE THICKNESS OF THE CORTEX IN BOTH SANE AND INSANE PERSONS AS DETERMINED BY DIFFERENT INVESTIGATORS. (*Donaldson.*)

	MALES.					FEMALES.					
Authority.	No. of Brains.	Defect.	Right Hemisphere.	Left Hemisphere	Average.	Average.	Left Hemisphere.	Right Hemisphere.	Defect.	No. of Brains.	
Conti ...	10	None	2·29	2·21	2·25	2·24	2·25	2·24	None	8	—
Franceschi	10	,,	2·479	2·474	2·48	2·46	2 457	2·463	,,	10	Normal
Donaldson	6	,,	2·91	2·94	2·92	2·91	2·92	2·89	,,	3	—
Major ...	4	Insanity	2·368	2·379	2.37	—	—	—	—	—	—
Bucknill & Tuke ...	35	Insanity	—	—	1·88	1·85	—	—	Insanity	30	Defective
Cionini	8	Gen. Paralysis	1·809	1·851	1·83	1·79	1·809	1·778	Gen. Paralysis	2	—

[1] Donaldson, *Am. Journ. of Psychology*, 1891.

In all probability these discrepancies are due to different methods of investigation, and different interpretations of the lower boundary of the cortical layer. At least these possible sources of error must be taken

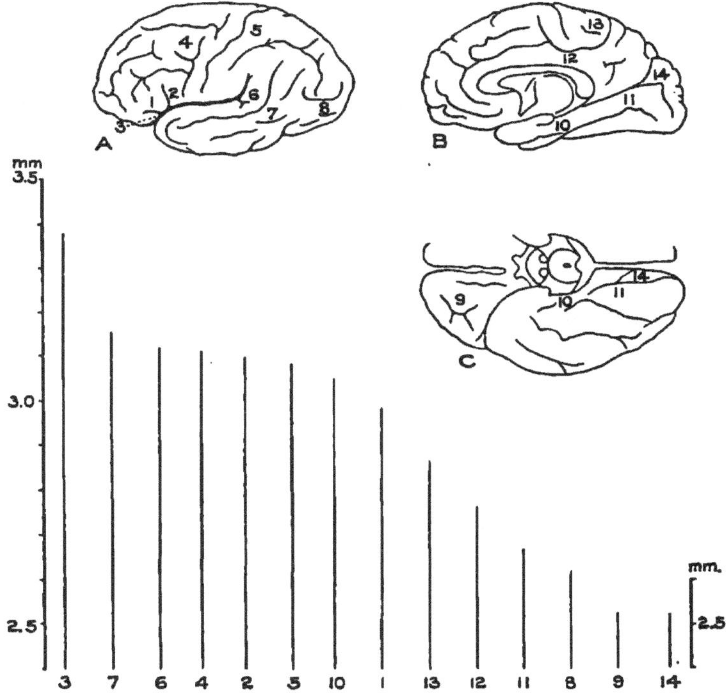

FIG. 39.—Illustrating the thickness of the human cerebral cortex. Averages from 9 cases: 6 males, 3 females. In each instance the thickness of cortex was obtained by averaging the thickness at the summit of the gyrus, at the side of the sulcus, this latter figure being taken twice, and at the bottom of the sulcus.

Only the ends of the verticals above 2·4 mm. are drawn. The numbers below the verticals show the localities of the cerebrum to which they apply (see A, B, and C); 3 refers to the middle of the insula, and the other numbers are self-explanatory; A, schema of the lateral view of the hemisphere; B, the mesal; C, the ventral. (Donaldson.)

into account before we attach any significance to the differences in question. The table indicates in the first place a thinner cortex in the cases of mental disease Among normal individuals my own figures, obtained from the so-called American brains, show the greatest thickness. Comparing the two hemispheres in the same brain series, it is found that the difference is extremely small, and that the figures hardly permit of the inference that either the right or the left hemisphere has normally the thicker cortex—a conclusion similar to that reached concerning the comparative weights of the two hemispheres. When, however, the average thickness in the brain of the males is compared with that of the females, there is in every instance a slight advantage in favour of the males. The constancy of this difference indicates that probably the cortex in the female brain is on the average thinner than that of the male; just as on the average the female brain weighs somewhat less than that of the male. I give here a curve (Fig. 39) which shows the average thickness of the cortex at the localities indicated on the accompanying outlines. In general the cortex is thickest on the convex aspect of the hemisphere, and thinnest on the ventral and mesal. It is also thicker in the middle of the hemispheres than at the poles.

The change in the architecture produced by growth and the significance of the constructional elements in the general arrangement of the central system form a separate line of investigation: to these matters the next chapter will be devoted.

CHAPTER XI.

ARRANGEMENT OF STRUCTURAL ELEMENTS.

Arrangement of spinal nerve roots—Meynert's scheme—Finer anatomy—Size of nerve elements—As related to size of animal—Mass of the cell—Mass of the nucleus—Innervation of the frog's leg—Gaule's observations on numerical relations—Connection of nerve elements with one another—Course of nerve impulse—Complexity of central system—Variations in complexity.

A CROSS-SECTION of the spinal cord like that given in Fig. 40 will show the arrangement of the afferent and efferent elements.

Such a section gives the structure of a modified segment of the cord, and since the entire central system is regarded as composed of such segments, more or less modified at the cephalic end, this structure may be regarded as typical. The afferent fibres arrive at the centre by way of the dorsal roots, and the efferent fibres leave it mainly by way of the ventral roots, but in small measure by the dorsal roots also. Toward the cephalic end of the cord and in the bulb the fibres carrying the efferent impulses, but which do not pass out by the ventral roots, become separated and form a lateral root intermediate between the other two. The central cells not shown in the figure are located in both the dorsal and the ventral zones of the developed cord. These cells are interpolated in the pathway of the incoming fibres, and serve to distribute the impulses which they receive to other localities within the cord or brain, thus increasing the possible number of pathways by which an incoming impulse may find passage. Since these cells are arranged in series, the impulses which take a long course must often pass through several groups of them.

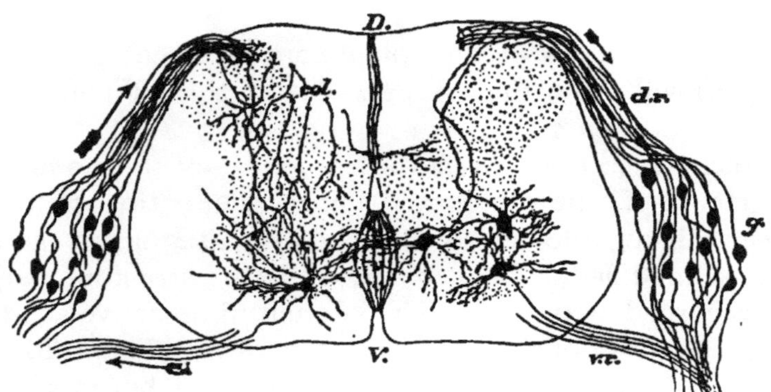

FIG. 40.—Cross-section of the spinal cord of the chick, × 100 diameters. (Van Gehuchten.) *D.*, dorsal surface ; *V.*, ventral surface ; *d.r.*, dorsal root ; *v.r.*, ventral root ; *g.*, spinal ganglion. On the left the arrows indicate the direction of the larger number of impulses. The small arrow on the right dorsal root calls attention to the fact that some neurons arising in the ventral plate emerge through the dorsal root and convey impulses in the directions indicated.

But concerning the maximum number of times that the impulse may thus be interrupted, no definite statement can be made. Physiologically, then, the entire central system may be regarded as made up of these afferent cells, over whose neuron the incoming impulses arrive ; of central cells which distribute them, and finally of a group of efferent cells, over the neurons of which they pass out. It is plain that, according to this scheme, the complexity of any given nervous system would depend more on the development of a group of central cells than on that of either of the other groups mentioned. There is, however, another and more formal way in which the structure of the central system can be presented. For the original description of this plan we are indebted to the Viennese anatomist, Meynert.[1] He looked upon the central nervous system as composed of three principal masses of grey matter—namely, the cortex of the hemispheres, the basal ganglia, and the central grey matter of the cord. (See Fig. 41.)

The fibres which connect the cortex (*C*) with the basal ganglia (*L* and *S*) form Meynert's projection system of the first order; those uniting the basal ganglia with the central grey matter of the cord, that of the second order; while the fibres (*D.r.* and *V.r.*) uniting the central grey matter of the cord with the periphery form the system of the third order. In addition, he distinguished two other systems of fibres within each principal subdivision of the central system: the commissural (*c c*), which connects symmetrical points on opposite sides of the median plane; and the associational (*a*), which connects points on the same side. An abundance of objections can be urged against the details of this scheme, and even some of the fundamental points which it was intended to illustrate must be abandoned, but it had in its time the value of a good working hypothesis. Applying to it the notion of the construction of the central system which I think more hopeful, it appears that the projection system of the third order includes our group of afferent and efferent fibres, and that the remnant of the scheme represents details of arrangement among the central cells, details which are most highly elaborated in the encephalon. The relations of

[1] Meynert, in Stricker's *Handbuch der Lehre von den Geweben*, 1872.

the parts of the nervous system as expressed in schemes such as the foregoing, suggest a further inquiry concern-

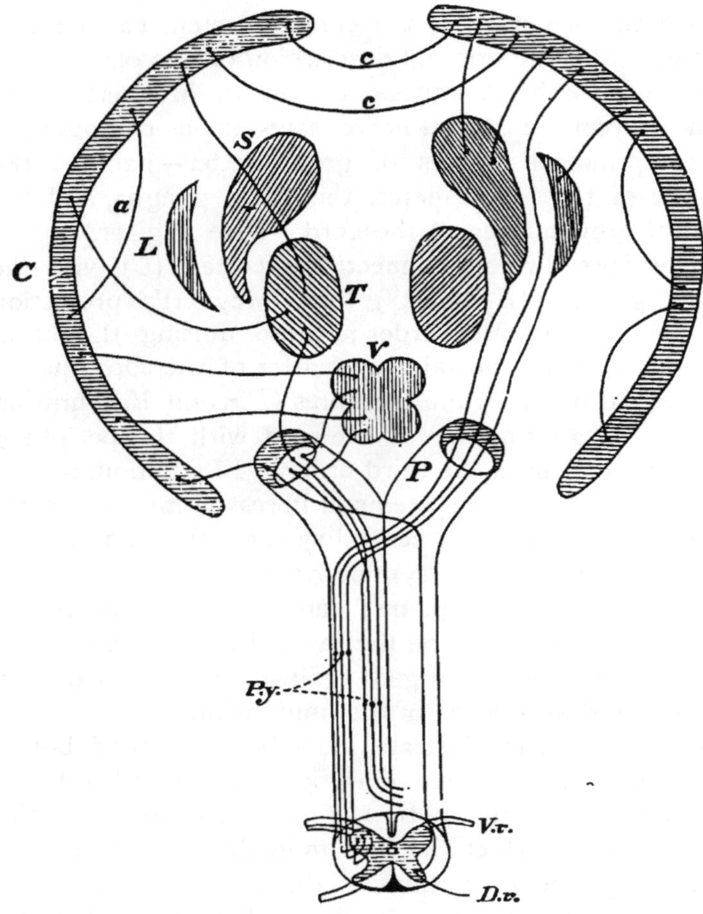

FIG. 41.—Schema of the arrangement of the principal masses of grey matter in central system, cerebellum being omitted. (Based on Meynert, from Landois, *Physiologie.*) *a*, association fibres; *C*, cerebral cortex; *c*, *c*, commissural fibres; *D.r.*, dorsal root; *L.*, lenticular nucleus; *P.*, pes; *Py.*, pyramidal tract; *S*, caudate nucleus; *T.*, optic thalamus; *V*, quadrigemina; *V.r.*, ventral root.

ing their elementary constituents. For the correct understanding of their form and mass this inquiry is indispensable, since in the absence of such facts an effort at interpretation might be fairly compared with an attempt to estimate the military power of two nations by weighing their armies while neglecting the fact of personal strength and courage, material equipment, degree of organisation, and the temper of the country.

As we know from the previous chapters, the nerve cells differ widely in the size to which they ultimately attain, but at the same time in different portions of the central system they exhibit peculiarities in shape and structure which are characteristic. It is important, therefore, to determine whether any general significance can be attached to these differences.

The relations between the diameter of the cell-body and that of the neuron arising from it have been previously expressed by the statement that it is correct to infer that a large neuron arose from a large cell-body, and this relation is useful when search is made for a group of cell-bodies to be associated with a group of fibres.

Most interesting on account of their wide application would be the facts bearing on the relations between the size of the nerve cell and the size of the animal, especially when within the same species, varieties widely different in size were compared. But such studies have still to be made. Strictly speaking, the observations alone available show a relation between the diameters of the cell-bodies and either the length or weight of the animals to which they belong. The size of the cells in the same region in animals of different species has also been studied. Table 51, taken from Kaiser, may be used as an illustration of this latter case. This author's studies were confined to the cells occupying the ventral horns in the cervical enlargement of the spinal cord. The different species of animals which he examined are named in the table. The cell-bodies were measured in two diameters, and the mean of these measurements is given in microns. The record in the table appears under two heads, that exhibiting the diameter of the chromophobic cell-bodies, or those staining poorly, and that of the chromophilic cell-bodies, or those staining well. This distinction is important because in the latter case the cell-bodies have the smaller diameter. In this locality, therefore, the chemical constitution varies with size.

TABLE 51.—SHOWING IN A SERIES OF MAMMALS THE MEAN DIAMETER OF CELLS FROM THE VENTRAL HORNS OF THE CERVICAL ENLARGEMENT. THE MEASUREMENTS FOR THE CHROMOPHOBIC AND CHROMOPHILIC CELLS ARE KEPT SEPARATE. (*Kaiser.*)

MEAN DIAMETER OF CELLS IN μ.			
Chromophobe Cells.		Chromophile Cells.	
1 Plecotus auritus ...	28–53		
2 Talpa Europæa ...	36–54	1 Talpa Europæa ...	17–40
		2 Erinaceus Europæus	25–45
3 Cercocebus sinicus...	33–60	3 Cercocebus sinicus ...	23–46
4 Cuniculus domesticus	41–61	4 Cuniculus domesticus	32–57·5
		5 Homo	23–59

1 Bat. 2 Mole. 3 Monkey. 4 Rabbit. 5 Man.

On passing from the bat up to man there is an increase in the diameter, but this increase is by no means proportional to that in the body-weight. While, therefore, the larger animals in this instance have slightly larger cell-bodies, there is no constant relationship between the diameters of the cell-bodies and the weight of the animals. We know, further, that the homologous nerve cells in the spinal cord of the horse and the ox have cell-bodies only slightly larger than those found in man. It is necessary, however, to be on guard against the somewhat misleading character of linear measurements. For example, a cell 53 μ in diameter has, when its volume is calculated as a sphere, only two-thirds the volume of one 59 μ in diameter. The slight difference in diameter is therefore quite significant when translated into volume.

In this description the expression cell-body has been reiterated for the sake of contrasting the measurements on the body with those which are next to be considered, and which relate to the mass of the entire cell. Supposing the cell-bodies were even of equal volume in the bat and in man, nevertheless owing to the enormously larger absolute size of the human nervous system, the neurons must have a much greater length in man than in the bat, making the mass of the entire cell many times greater. The converse is true of the largest mammalia when their nerve elements are compared with those of man, but just how different the relative development of the nerve cells in the different mammals may be is still undetermined. Formulating these results, they may be expressed as follows: In the mammalian series cell-bodies do not regularly increase in weight with the increase in the body-weight. On the other hand, the mass of the neuron is more closely correlated with this body-weight, since the larger the animal the greater the length as well as the diameter of the neuron. This latter relation is suggestive, for it follows that in the large animals a very much greater mass of nerve substances, represented by the neuron, is under the control of a single nucleus. At the same time it is not possible to see that there is any other striking difference between the two nerve cells, unlike in the mass of their neurons, except that the nerves impulse has a shorter path in the shorter neuron. It must, however, be left undecided whether a large neuron is either an advantage to the nerve cell by adding to the quantity of cytoplasm which can store energy, or a disadvantage as representing an outgrowth which requires to be maintained. Perhaps future experiments will gain something in definiteness if in measurements the comparison be made not alone between the cell-bodies, but between the volume of these and also of the entire cell and the nucleus.

Mason has studied the size of the nucleus in the frog and certain reptiles.[1] His results indicate among the nuclei relations similar to those found for the cell-bodies of mammals examined by Kaiser. The animals represented in his table had widely different body-weights, nevertheless there is no correlation between their body-weights and the diameters of the nuclei as there described. The explanation which Mason proposes is based on the connection of the larger nuclei with the cells controlling the most active and most bulky muscles. Scattered observations on the nerve cells of mammals indicate similar relations.

Physiological reactions and the mass of nerve elements are interdependent. Gaskell has emphasised the relation between the diameter[2] of the fibres and their function, when describing the bundle of extremely fine fibres which in mammals passes out by the ventral roots of many spinal nerves, and which he was able to trace into the sympathetic system. It has long been a familiar fact, also, that the different tracts within the spinal cord are characterised by fibres of different calibre, but in order to interpret this it is necessary to compare

[1] Mason, *Journ. of Nerv. and Ment. Dis.*, 1880, 1881.
[2] Gaskell, *Journ. of Physiol.*, vol. vii., 1885.

fibres of different sizes having the similar physiological functions.

I have made the attempt to determine whether the number of nerve fibres going to the segment of the frog's leg above the knee as compared with the number going to the segments below the knee bore any relation to the mass of muscle and area of skin belonging to these divisions, and also whether the nerve fibres were of different diameters according to their destination. The frog (*Rana catesbiana*) which was used weighed 13·5 grammes, the muscles of the entire leg weighed 1·9 grammes, of which 1·3 grammes represented those above the knee, and ·6 those below, a ratio of 2·16–1. The total area of the skin on the leg was 21·3 sq. cm., being 6·5 sq. cm. above, and 14·8 sq. cm. below, the knee, a ratio of 1–2·2, or almost exactly the reverse of that found for the weight of muscle. If the area of the cross-section of the nerve trunk were proportional to the weight of muscle and area of skin taken together, it might be expected that it would be nearly the same for the two portions of the limb. The following table exhibits the relations observed:—

TABLE 52.—SHOWING IN A FROG OF 13·5 GRAMMES THE AREA OF THE PORTION OF THE NERVE TRUNK, THE NUMBER OF FIBRES, AND THE AVERAGE AREA OF THE FIBRES, DISTRIBUTED IN THE ONE CASE ABOVE THE KNEE, AND IN THE OTHER, BELOW IT. ALSO THE RELATIVE WEIGHTS OF THE MUSCLE MASSES AND THE RELATIVE AREAS OF SKIN IN THESE TWO PORTIONS OF THE LEG.

MUSCLES AND SKIN OF FROG'S LEG.	AREA OF THAT FRACTION OF THE NERVE TRUNK.	NUMBER OF FIBRES.	AVERAGE AREA OF SINGLE FIBRES.	RELATIVE WEIGHT OF MUSCLE.	RELATIVE AREA OF SKIN.
Above knee	·27 mm.^{2}	1577	171 μ^2	2·16	1
Below knee	·13 mm.^{2}	1949	67 μ^2	1	2·22

The area of the cross-section, therefore, of the nerve trunk which supplies the muscles and skin above the knee is in the ratio of 2·07–1 to the area of nerve supplying those below it. The area of the trunk for the parts above the knee is thus more than twice that for those below. Why?

By enumerating the fibres which the two trunks contain, it is seen that the larger nerve is composed of 1,577 fibres, while the smaller contains 1,949. The nerve trunk of largest size contains, therefore, the largest fibres, and it is the size, not the number of them, which explains its area. To interpret the foregoing data it should be remembered that the sensory and motor nerves are not equal in number. From observations made by Birge, it is calculated that in the nerve roots making up the sciatic nerve the number of sensory fibres is to the motor in the proportion of 2·7–1. Taking this as a basis, and assuming that the number of nerve fibres in the two divisions of the leg is proportional to the weight of the muscles and the area of the skin, we find that there would be 1,451 fibres above the knee and 2,075 below, figures which are near to those determined by direct enumeration. According to this enumeration, the portion of leg below the knee is controlled by a greater number of cells than the portion above it, but the fibres passing below the knee are smaller in average diameter. The interpretation of the anatomical relations of the nerve supply to the hind leg of the frog is, therefore, as follows: The number of efferent nerve elements is roughly proportional to the weight of the muscles. The cells which innervate the muscles above the knee have large cell-bodies and large neurons, but they are directly associated with only a small number of sensory nerves. The control of these muscles is therefore strong and persistent, but coarse. In the muscles below the knee the condition is reversed. The efferent cell-bodies are small, the neurons small, the sensory control very large, and the reactions less strong and less persistent, but more delicate. According to this view, the refinements of the reactions depend on the relations between the motor nerves and the sensory group associated with them, either directly or through the agency of central cells. It so happens that in the case of the nerve supply to the limb these cell elements of large size are associated with a group of sensory cells small in number, but this is not a necessary combination. Taken alone, then, we interpret size in nerve elements as the anatomical expression of potential energy.

Turning to the central nervous system, it is seen that, contrary to the arrangement that appears in the innervations of the limbs, the fibres which in the pyramidal tracts pass from the cortex to the lumbar enlargement of the cord have the longest course and also the greatest average diameter. The length of the course here is not so important as the fact that the great number of cells in the lumbar enlargement is put under the control of the cortex by the aid of a few fibres. The control is not refined or delicate in this case, and the anatomical expression of this lack of refinement is found in the association of each cortical cell with a large number of motor cells in the cord. Size here is associated with coarseness of organisation. It has been stated by several writers that the largest fibres had the longest course in the peripheral nerves, and hence that increase in the diameter of the nerve fibres probably facilitated the passage of the nerve impulse in much the same way that increase in the diameter of an electric wire facilitated the passage of an electric current. The large fibres were, however, not to be found at the periphery. To explain the absence of fibres of large diameter from the distal portions of the peripheral nerve it has been suggested that the neuron undergoes a conical diminution, so that its cross-section becomes gradually smaller as it passes away from the central system (Schwalbe).[1] Since in the frog, at least, the largest neurons can be followed into the branches which are given off to the different muscles of the thigh, a further explanation of their absence from the main trunks at the periphery appears unnecessary.

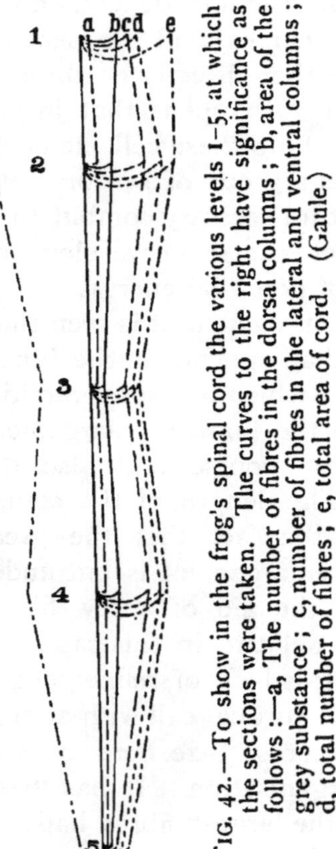

FIG. 42.—To show in the frog's spinal cord the various levels 1–5, at which the sections were taken. The curves to the right have significance as follows:—a, The number of fibres in the dorsal columns; b, area of the grey substance; c, number of fibres in the lateral and ventral columns; d, total number of fibres; e, total area of cord. (Gaule.)

These observations on the innervation of the frog's leg harmonise with the observations of Ross[2] and of Kaiser to the effect that it is the smaller cells, and those lying at the periphery of the various cell groups in the spinal cord, which send their neurons to the more distal segments of the limb. These cells, developing late, are known to have bodies of small size, and since it is found that the neurons passing to the more distal segments of the limb are small in diameter, there is nothing incongruous in connecting such neurons with such cell-bodies.

[1] Schwalbe, *Ueber die Kaliberverhältnisse der Nervenfasern*, Leipzig, 1882.

[2] Ross, *The Diseases of the Nervous System*, New York, 1883.

Numerical relations in the architecture of the central system are still to be examined, and to these we next turn. Under the title of "The number and distribution of the medullated fibres in the spinal cord of the frog," Gaule[1] has published some interesting observations. His view of the animal organism, metaphorically expressed, is that of a complex organic molecule. In such a molecule the cell elements of the different tissues are assumed to stand related to one another as do the atoms in the molecule of the chemist. This conception involves the idea that a given number of one group of elements demands a corresponding number of another group. The application of this notion to the spinal cord of the frog is as follows: A certain number of afferent and efferent fibres connected with the cord requires within it a corresponding number of cells and fibres, which shall put the segmental elements in connection with one another and with the higher centres. From the previous investigations of Birge there were records of the number of nerve fibres in the spinal nerve roots of the frog, and these records form one basis for the subsequent calculations. According to Gaule's theory each root fibre, both ventral and dorsal, has usually eleven connections, which in certain localities determined upon may be reduced to a smaller number. By this expression is meant that in the cross-section of the spinal cord there will be found eleven fibres for each fibre found in the roots. To test this hypothesis Gaule made cross-sections of the spinal cord; (1) just at the point where it passes into the bulb; (2) through the middle of the cervical enlargement; (3) in the upper half of the thoracic region; (4) through the upper portion of the lumbar enlargement; (5) through the

[1] Gaule, *Abhandl. d. Mathemat.-Physiol. Cl. Königl. Sächs Gesellschaft der Wissenschaften*, 1889.

lower end of the lumbar enlargement. He then counted all the fibres in these sections, making the table for these localities, given below, and putting first the number of fibres in the cross-section of the cord for which the hypothesis called, and then the number which was actually enumerated, the results were found to be in close agreement.

TABLE 53.—SHOWING THE NUMBER OF MEDULLATED FIBRES IN THE CROSS-SECTIONS OF THE SPINAL CORD AT THE LEVELS NAMED: FIRST, ACCORDING TO HYPOTHESIS; SECOND, ACCORDING TO ENUMERATION. THE LEVELS CAN BE LOCATED IN FIG. 42. (*Gaule.*)

LEVEL OF SECTION.	DEMANDED BY HYPOTHESIS.	FOUND BY ENUMERATION.
1	56,000	56,674
2	74,000	74,699
3	45,500	41,825
4	60,500	61,058
5	18,000	16,313

A great many objections have been urged against these figures, nevertheless there is here indicated an orderliness in construction which was by no means suspected, and which is of fundamental importance in further studying the architecture of nerve centres. It has been urged that the neurons divide in their course. A few such observations have been made, but in the peripheral nerves of man and the frog these divisions of the neuron are exceptional. In the spinal cord of fish such a division is seen in the case of the giant fibre of Mauther, and the same also has been observed in the frog, while recently Sherrington has reported "geminal fibres" in man. But at best the number of such geminal fibres is small. When, however, the branches of the dorsal root fibres after they enter the cord, or the branches given off laterally from the neuron just after its origin from the cell-body, are taken into consideration the case is quite altered. These branches are numerous, and probably some of them become medullated, and hence within the central system the idea of a one-to-one relationship between cell-bodies and neurons breaks down entirely. The relation, however, which Gaule sought to establish was between the number of root fibres and the number of medullated fibres in the cross-section of the cord. The figures bearing on this would a few years hence have been interpreted to mean that the medullated fibres within the cord each stood for a single cell, and though at present that is not warranted, it is the interpretation of Gaule's facts, and not the facts themselves, which are thereby affected.

The many branches of the cell-body form channels by which it becomes associated with other cells, hence the manner in which cell elements are related by means of their outgrowths is thus brought to our attention. The older histologists held that the nerve cells and nerve fibres of different origin were often structurally continuous, so that one fibre might have similar connections with two cell-bodies. Cells joined to one another by broad commissural outgrowths were often pictured, and the conception of the nervous system was that of a very extensive network, in which at least large numbers of the elements, and perhaps all, were involved. According to the modern view, based on embryology, a neuroblast gives rise not only to the cell-body, but to all its numerous prolongations. These can be followed into the neighbourhood of other cell-bodies and prolongations, but as a rule continuity between two structural units is not described. In the spinal cord of some of the electric fishes (Fritsch) the cell-bodies, united by broad protoplasmic bands, can be seen, but the developmental history of this arrangement has never been followed, and in the light of our present knowledge is to be looked upon as exceptional.

Granting the lack of anatomical continuity, the ques-

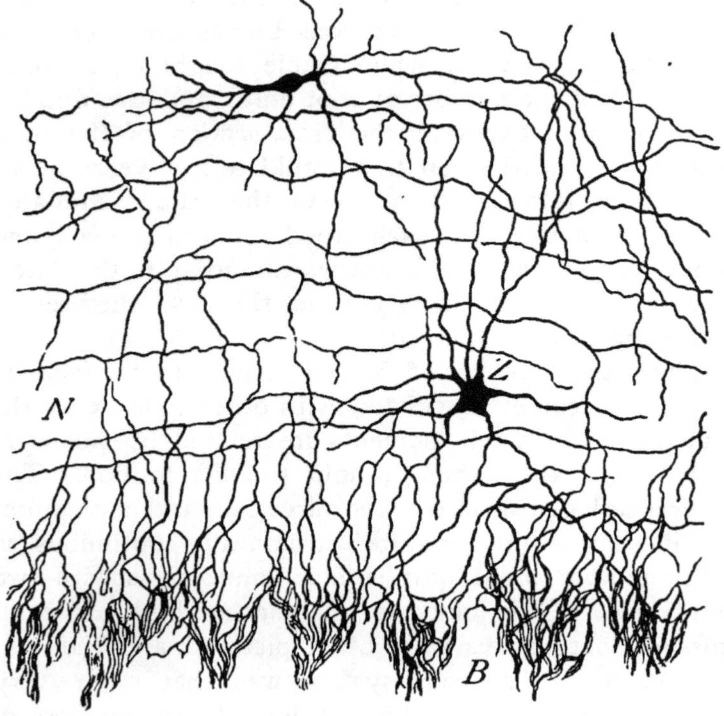

FIG. 43.—Showing at the lower edge of the figure a series of basket-like terminations of neurons which surround the bodies of the great cells of Purkinje in the cortex of the cerebellum. Ramón y Cajal.) *Z*, cell-body; *N*, neurons; *B*, basket-like terminations arising from cell *Z*, and enclosing the cells of Purkinje.

tion arises how the physiological reactions of the nervous system can be explained without it. By way of illustration, it may be urged that there is apparently no difficulty in the passage of the nervous impulse from the motor nerve to the muscle which it controls, although the connection here is secondary. In the same way the sensory epithelial cells arouse impulses in the fibres with which they are associated, although again the connection is not genetic. Here and there in the central system nerve fibres are found, apparently terminating in small disc-like expansions on the surfaces of cells. In some cases the nerve fibre terminates in an elaborate brush, which forms an enclosing basket about the body of a second cell. In other cases the neuron forms its final brush in the neighbourhood of the neuron, or the dendrons of another cell.

On the other hand, it is not easy to make out a close relation between the branches of the different cells, as they interweave with one another. In many cases where it is assumed that the impulse passes from a terminal brush of a neuron into the branches of another cell, the specimens do not show continuity between the two. The difficulty, therefore, is to conceive the anatomical relations, so that the impulses may not be stopped by interruptions in their course.

It must be assumed either that these various branches are structurally continuous with one another, although the continuity is hard to demonstrate; or that contiguity of these structures is sufficient, the impulse in some way passing across the interval; but whichever view is taken, the actual relations, as shown in sections, require to be largely completed by hypotheses. As the matter now stands, however, the greater weight of evidence is in favour of the notion of contiguity.

Many problems are bound up in this general question. For example, an incoming impulse reaches the nervous system by one of the dorsal root fibres. These branch immediately after entering the spinal cord, one branch passing cephalad and the other caudad.

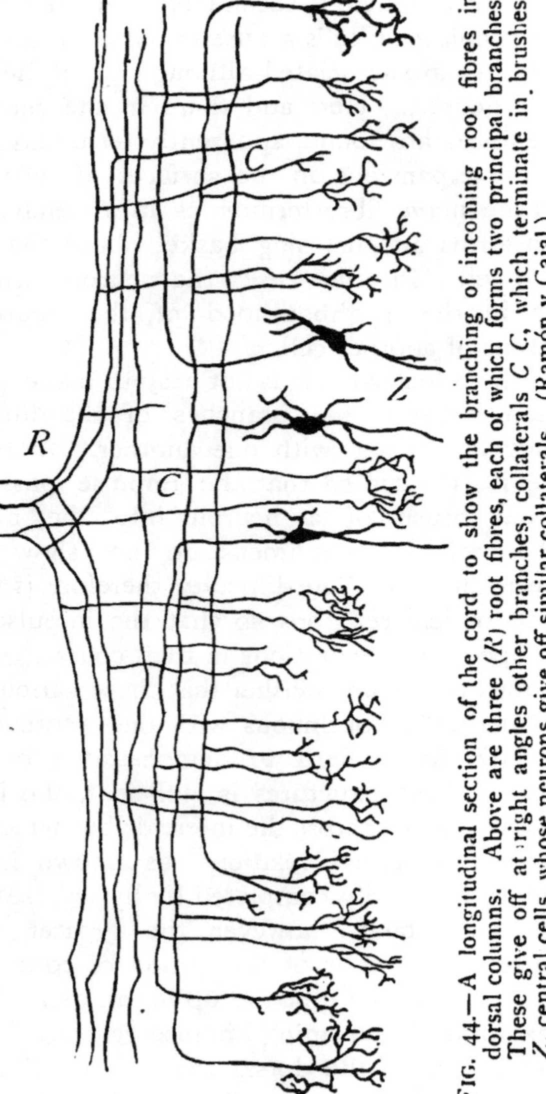

FIG. 44.—A longitudinal section of the cord to show the branching of incoming root fibres in dorsal columns. Above are three (*R*) root fibres, each of which forms two principal branches. These give off at right angles other branches, collaterals *C C*, which terminate in brushes. *Z*, central cells, whose neurons give off similar collaterals. (Ramón y Cajal.)

From these main branches are given off smaller ones, which penetrate the grey matter. These collateral branches, as they are called, may be very numerous, and we desire to know whether in each case the impulse that comes in by the main stem is distributed in the cord through all these branching pathways. Certainly different reactions follow the stimuli which must arrive at the cord by the same fibres, and it is difficult, therefore, on the one hand, to see how there should be a choice of pathways open to the impulse; and, on the other hand, how, if such choice does not exist, successive reactions should not be more alike.

Possibly some basis for an explanation can be found in the fact that the series of cells to which these collaterals deliver the impulse may not on the several occasions be in the same condition, or that the impulse may be here and there blocked on reaching the interval separating the cell branches, and so at different times different groups of cells respond with variations both in order and intensity. The following example may serve as an illustration of this hypothesis. There is reason to think that a given branch or collateral may terminate in the neighbourhood of several cells in such a manner that it is able to deliver an impulse to all of them. If the next step in the physiological process depends upon the condition of the receiving cells, the greater their number and the more open they are to stimuli from other sources the more diverse would be the responses. It thus happens that the machine-like reactions of the nervous system would be best carried out by a small number of nerve elements holding a highly constant relation to one another and acted on by a minimal number of modifying stimuli, while a larger number of elements and more numerous avenues for afferent impulses would permit of greater variations in the reactions. Objectively, one measure of central complexity is the modification of the outgoing impulse in response to slight variations in the stimuli. Therefore we should expect that a nervous system composed of a large number of structural elements, well organised, would give more highly modified reactions than one which contained a smaller number though similarly organised.

To make plain this view, it must be understood that not only are stimuli received through the sense organ which initiates a reaction, but we are all the time receiving concomitant stimuli through the other organs of sense, and as each group of responsive cells or those initiating the outgoing impulses can be influenced in some degree by any of the incoming impulses, it is evident at a glance that an almost endless series of combinations is possible. For example, a sharp cry may instinctively lead us to draw back from its source. If, however, the visual impression that accompanies the cry arouses sympathy, we draw towards it. The motor reaction in the two cases are opposites to one another, and they are modified by a secondary visual impression accompanying the auditory one. This is a very simple case, but it is an easy matter to imagine the various complications, when it is also remembered that any other sense impression, present or past, may enter in as a modifying circumstance. To regard, therefore, the nervous system simply as a mechanism capable of giving very complex reactions in response to stimuli, implies no more than the acceptance of the every-day point of view, taken when we venture to predict the actions of our neighbour under an anticipated set of conditions. It appears that the more complicated the nervous system of an animal is, the greater is the possibility of those refined adjustments in which several of the sensory elements shall play their proper parts, and each element controlling the outgoing impulses shall respond in due order and with tempered force.

Taking the nervous system as a whole, and for the moment neglecting the manner in which its constituent elements are grouped, it has been shown that the structural elements do not increase in size in proportion to the size of the animal. Yet the spinal cord of a cat would not suffice for a lion, nor that of a mouse for a beaver, for in the cords of the larger animals, since the individual cell-bodies are not so much larger, there must be absolutely a greater number of cell elements as well as great extension of the neuron to make up their bulk. Not only is there an increase in the number of elements representing the spinal nerves, but also in that of the central cells and in any such series, the possible combinations. theoretically increase far more rapidly than the absolute number of the elements. These central combinations rapidly increase with the increasing number of the central elements, though the number of them is indeterminate, and probably variable. Nevertheless, growth contributes to central complexity and organisation, and something will be gained by regarding the matter from this side.

CHAPTER XII.

ARCHITECTURAL CHANGES DUE TO GROWTH.

Causes of change due to growth—Size and organisation not neces sarily connected—Polarity of cells—Significance of neuron—Meaning of medullation—Formation of gyri—Medullation in cerebellum — Growth of cerebellar cortex—Development of human cortex—Source of cells in developed cerebral cortex—Observations by Vulpius—Observations by Kaes.

FROM what has been stated already, it is evident that the architecture of the nervous system in the young must be different from that in the adult, yet thus far the structure at maturity has alone been described. The possible causes of the changes due to growth may be classed under four heads: alterations in the number of the elements; their size; their organisation; and their nutrition. Since the central system is here considered only after the total number of cell elements has been formed, no variation due to this factor occurs. On the other hand, the alterations in nutrition due to age are so inadequately known, that in the absence of other evidence, nutritional conditions from the standpoint of the individual cell must be assumed to grow steadily worse from birth on. There remain, therefore, as important causes, the variations in the size and shape of the elements, together with those changes in their relation to one another which lead to differences in organisation.

On examining the developing system, the neuroblasts are found to be spherical and possessed of one process only, the neuron. It would be hasty to infer that elements in this stage of development were unorganised, since a central system apparently in this condition can give complex reactions.[1] In seeking an explanation for such a case, it should be remembered that if the impulses pass between fully-developed nerve elements by means of their contiguous outgrowths, then in the earlier stages of growth the contiguity of cell-bodies themselves might furnish a sufficient though less delicate arrangement for their transmission. In the mature system, however, organisation depends on development of the cell outgrowths. The most important of these, the neuron, is the first to appear; later, the dendrons. By means of the former, distant connections are established if it so happens that the neuron pursues a long course as a main stem, having but a few lateral branches near its origin. If, on the other hand, the neuron at once breaks up into many branches in the neighbourhood of the cell-body, these, like the lateral branches of the main stem, are looked upon as the basis of local connections.

Not only ontogenetically, but phylogenetically also, the dendrons appear latest. They must, therefore, be considered as important to the better organisation of the central system. That changes in the size and form of the elements usually accompany organisation is amply evident, though these changes are not necessarily correlated, for a system composed of small cells may yet be well organised.

It has been so long customary to describe the nerve cells without much analysis of them, that a number of peculiarities in the individual elements have but recently

[1] His, *Rep. 10th Internat. Med. Cong.*, 1890.

received proper attention. Among these peculiarities is a modification of portions of the cytoplasm for the production of the neuron. This differentiation is spoken of as polarisation.

It is probable that all neuroblasts are polarised so that the end of a mononeuric cell furthest from the primitive surface of the body is that from which the neuron grows out, its direction being away from this surface. These relations in the dineuric cells have not been determined. During the development of a mammalian nervous system, many foldings occur, but, provided the cells remain fixed, their polarity is preserved. Some of the neuroblasts, however, migrate, and the relative position of other cells changes in ways unknown.

Taking all these facts into consideration, it is not remarkable that now and then a neuroblast accomplishes but incompletely the series of changes marked out for it, and its neuron either develops imperfectly or fails to take the proper direction. It is thus possible that increase in size in a given encephalon may go on almost normally, helped in some case, to be sure, by an increase in the supporting tissues, but without the accompanying organisation, a phenomenon illustrated by the high weight of the brains of some idiots. It is thus seen that the weight and organisation of the encephalon are not necessarily correlated, a fact which must modify all inferences made directly from encephalic weight to intelligence.

Since the most important and at the same time the most distant connections of a cell depend on the neuron, it may be considered in more detail. This prolongation of the cytoplasm forming the axis cylinder of the nerve fibre is usually surrounded by a medullary sheath, a secondary structure, the mass of which has already been discussed, and found to be correlated with that of the axis in such a manner that the larger axes have the thicker sheaths. A physiological connection between the two is thus suggested, and it has been shown that with the acquisition of the myeline the nerve fibres become functional, a fact which has been effectively used to determine the order in which the several parts of the central system are organised. Medullation begins in the cord and spreads to the encephalon, the local mechanisms in the cord being functional before the connections between them and the higher centres are completed. In general, the elements first developed become largest in size, and since the most necessary functions presumably are those for which provision is earliest made, the relative size of the nerve elements in a given animal points to the degree of their physiological value.

In the central system the distance to be traversed by the neurons is an important factor. It appears that the long distance connections tend to be established early when the low specific gravity, the paucity of supporting tissues, and the small absolute size of the entire system, all combine to offer the least obstruction to changes of this nature. The associations later formed become predominantly more local with advancing age, consequently the axis cylinder which is to make a long distance connection early spans the greater parts of the interval between the centres that are to be associated. Later, not only is the absolute length greatly increased, but it approaches nearer the elements which it is destined to control. The first arrangement may be considered as a sketch plan of the adult system, and for the reason just given the further the fœtus departs from the first stages of its development the less feasible is any alteration in this framework. Yet the outlines may become heavier, and the few neurons which in the first instance form them may be later increased by others following them as guides, in much the same manner that the regenerating peripheral nerve follows by the aid of the old sheaths its previous course down a limb, although quite unable to force a new passage. From the observations of Cajal it appears that the lengthening neuron is enlarged at its tip in a way that suggests the comparison of it with a growing rootlet.

Often the axis cylinder acquires a great length before medullation occurs, and experiments showing that functional activity accompanies or immediately follows the formation of that sheath, have led to the view that the sheath is necessary for the activity of the neuron. It is not possible to make general statements concerning the significance of the sheath, for the reason that there are fibres which are functional, but in which the sheath is permanently lacking, and also because even in the typical medullated neuron there is one stretch of the axis just after it leaves the cell-body, and others at its final termination, where no sheath is found. It therefore is more in harmony with the facts to consider that in those neurons usually medullated the establishment of the final relations which makes the transfer of the nerve impulses possible, is the prime event, and that this is accompanied by the physiological change leading to medullation, a change which in these cases is an index of organisation. For the absence of this sheath in certain classes of nerves, and in portions of others, no explanation has been offered..

In one way the final approximation of the cell outgrowths, accompanied by medullation, is the most important condition modifying brain architecture, but it is by no means so evident as the grosser changes which accompany it.

Preceding medullation for the most part, none among

these changes is more striking than that produced by the formation of the gyri of the cerebral hemispheres. Up to the fifth week of fœtal life the hemispheres are smooth. Then appear the transitory fissures, caused by the infolding of the entire thickness of the wall of the hemispheres, a process which has been attributed by Cunningham[1] to inequality in growth caused by the over-rapid enlargement of the encephalon as compared with that of the cranium. This inequality is soon relieved by proportionate expansion of the cranial bones, and almost all the transitory fissures disappear, one, or perhaps two, remaining. The next formation of gyri, which begins about the twentieth week with the appearance of the central fissure, does not involve the entire thickness of the hemispheral wall, but is due to a retardation of growth along the line of the future fissures. In the growing cerebrum both the bottoms of the fissures, as well as the summits of the gyri which limit them, move away from the geometric centre of the brain, the latter more rapidly, and hence the fissures first formed become gradually deeper so long as the brain continues to enlarge. Variations in the rate at which the radii of the several portions of the cortex elongate become more numerous as the brain grows larger, so the number and extent of the fissures is also increased. The causes of this change are but poorly understood. Both the radial fibres that grow into the cortex, as well as those that grow out from it, must have an influence, and probably an important one, since they are the fibres by which it is most affected, while the various sorts of non-radial fibres are of hardly less significance. One effect of this fissuration is to increase the surface of the cortex, a change which must be accompanied by a

[1] Cunningham, *A Contribution to the Surface Anatomy of the Cerebral Hemispheres*, Dublin, 1892.

lengthening of all the fibre systems that run parallel with its surface.

These conditions being complex, it follows that the character of the fissuration may vary greatly in consequence of a slight alteration in them, and nearly the same gross appearance be the result of dissimilar combinations among them. For instance, a much-fissured surface may be the result either of an ample

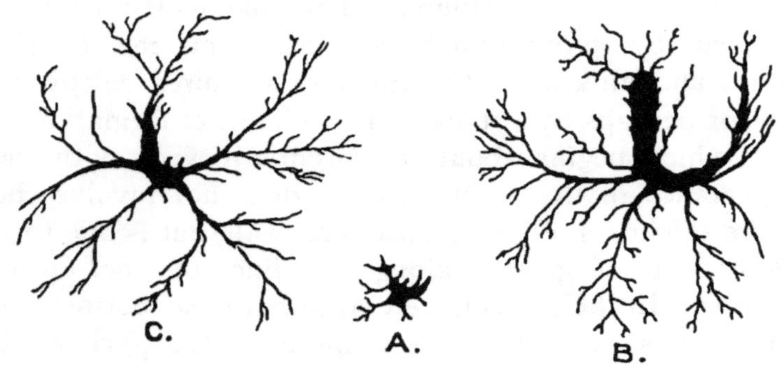

FIG. 45.—Silhouettes of the white substance (arbor vitæ) exposed by a longitudinal median section of the human cerebellum. A, at birth ; B, type in males and in vigorous and muscular persons ; C, type in females and in persons weak and aged. Natural size. (Engel.)

expansion of the cortical layer, or a lack in the formation of the subcortical fibres.

The fibres in the encephalon become only slowly medullated, so that from birth to maturity there is an increase of the portion medullated. Engel[1] has shown this in the case of the cerebellum. When a section in the median plane is made through the cerebellum there is exposed a central mass of white matter, having many branches, and surrounded by the grey cortex. This "arbor vitæ," as it has been called, from its resemblance to the foliage of the tree bearing that name, changes in shape, and at

[1] Engel, *Wien. Med. Wochen.*, 1863.

maturity acquires longer and more numerous branches, which may be lost again under pathological conditions. In Fig. 45 are given the outlines of the arbor vitæ as observed by Engel.

The cortical layers both in the cerebrum and cerebellum grow thicker as the brain increases in size. The studies of Krohn on the cerebellar cortex of the cat show the molecular layer to be thicker in the cat of six months than in the kitten of three months.

TABLE 54.—SHOWING THE THICKNESS OF THE MOLECULAR LAYER OF THE CEREBELLAR CORTEX IN CATS AGED RESPECTIVELY THREE MONTHS AND SIX MONTHS. MEASUREMENTS IN MILLIMETERS.

It will also be noted that this layer is thicker on the left side. This is constant. (*Condensed from Krohn.*)

AGE.	RIGHT HALF.	LEFT HALF.
3 months	·278	·295
6 months	·321	·345

Probably the thickness of the cerebral cortex increases so long as any of its layers continue to grow in thickness, and there is reason to think that this may go on up to the fortieth year. The development of the cortex is very instructive, and is illustrated by the accompanying figures, based on those by Vignal.[1]

In the fœtus of the twenty-fourth week small granule-like cells, arranged in vertical rows and closely packed together, form the bulk of this layer. In the twenty-eighth week it has increased in thickness, the cells have

[1] Vignal, *Développment des Éléments du Système nerveux cérébro-spinal*, Paris, 1889.

become somewhat stratified, and in the lower portion of the layer (A, III) have begun to show the full characters of nerve cells, a nucleus, nucleolus, and well-marked cytoplasm. A month later (B) cells have begun to develop through the entire thickness of the cortex, the clearly marked cells have become larger, and the elements have further separated. The failure of the cortex to increase in thickness during this time may be associated with the fissuration of the hemispheres which causes an extension of the surface. At birth the thickness is much increased, more cells are developed, and those previously enlarged have increased in size. Between this and maturity an increase in thickness is not shown in the figures, but equally important is the change in the extension of the cortex, which at birth covers an encephalon weighing only 382 grammes, to an area which will cover one weighing 1,350 grammes; added to which is the further extension demanded by the deepening of the fissures, since the entire area at maturity amounts to more than three times that found at birth. It is no wonder, then, that the cortical elements become still further separated.

For the cells continually appearing in the developing cortex no other source is known than the nuclei or granules found there in its earliest stages. These elements are metamorphosed neuroblasts which have shrunken to a volume less than that which they had at first, and which remain small until, in the subsequent process of enlargement necessary for their full development, they expand into well-marked cells. Elements intermediate between these granules and the fully developed cells are always found, even in mature brains, and therefore it is inferred that the latter are derived from the former. The appearances there lead also to the conclusions that many elements stop short of com-

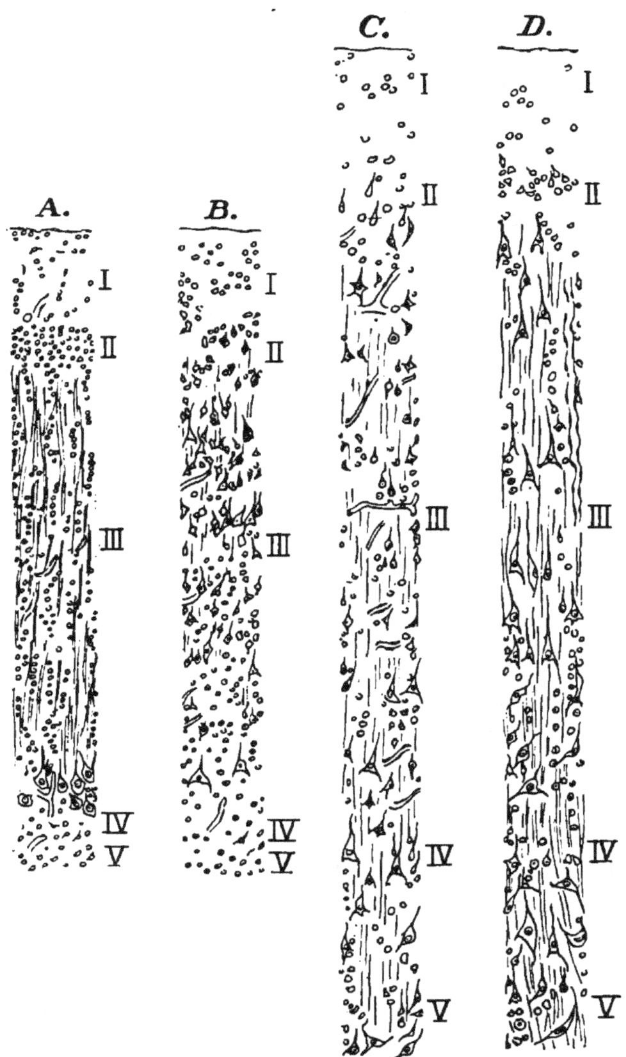

FIG. 46.—To show the developing human cortex (Vignal), × 40 diameters. *A*, fœtus of 28 weeks; *B*, fœtus of 32 weeks; *C*, child at birth; *D*, man at maturity; I–V, layers of the cortex according to the enumeration of Meynert.

plete development, that the number of elements which might possibly develop in any given case is far beyond the number that actually does so, and that the characteristic appearance of the cortex in the various localities depends in a measure on the expansion of dissimilar layers of the primitive granules.

Just here can be met a difficulty sometimes felt in the explanation of these changes. It has been urged that the possible number of cells latent and functional in the central system is early fixed. At any age this number is accordingly represented by the granules as well as by the cells which have already undergone further development. During growth the proportion of developed cells increases, and sometimes, owing to the failure to recognise potential nerve cells in the granules, the impression is carried away that this increase implies the formation of new elements. As has been shown, such is not the case.

These conclusions require to be tested further, but bearing on them is the fact that the number of granules and partially developed cells was excessive in the defective portions of the cortex of the blind deaf-mute, Laura Bridgman, in whom normal development in these localities ceased at the end of the second year of life (Donaldson), and similar appearances have been found in the cortex of persons congenitally defective. Applying these ideas to the comparative anatomy of the cortex, there is reason to think that in passing down the zoological scale the proportion of undeveloped elements would increase, and for this there is positive evidence. It is conceivable, too, that in the case of man wide individual variations in the proportion of developed granules may occur, and with it would come corresponding differences in the organisation of the cerebrum, without much alteration in its absolute weight. If the history of the cortex up to maturity is followed with care, changes of great interest are to be discerned, and these changes are apparently open to much individual variation. Their importance is enhanced by the fact that on the one hand the cortex is the most complicated group of the central cells by which cross references between incoming and outgoing impulses are established, and on the other by the fact that the changes in question have been shown to occur during the period of formal education, and may therefore be in some measure influenced by it. The accompanying diagram represents on one side in a schematic manner the distribution of the cells in the occipital region of the human cortex, and on the other the arrangement of the fibres. The facts which are to be presented bear on the relative development of these layers of fibres as indicated by their increasing medullation (Fig 47).

The change with age in the number of tangential medullated fibres or those running parallel with the cortical surface has been studied by Vulpius.[1] For the purpose of this investigation, the cortex was divided into three layers, an outer (*A*, *B*, *C* and outer half of *D*), middle (*D*, inner half), and inner (*E* to *H* inclusive), represented respectively by the layers I, II ; III, and IV, to VIII in Fig. 47. The proportional number of the fibres was determined by counting them in a small area. Six localities in a large number of brains ranging in age between the thirty-second week of fœtal life and seventy-nine years were studied. The results are partially summarised in the two figures which follow, in which is recorded the proportional development of tangential fibres under the conditions named. The figures show that between the age of sixteen months and thirty-three years the number of fibres in all the

[1] Vulpius, *Archiv. f. Psychiat. u. Nervenkrank.* Bd. xxiii. 1892.

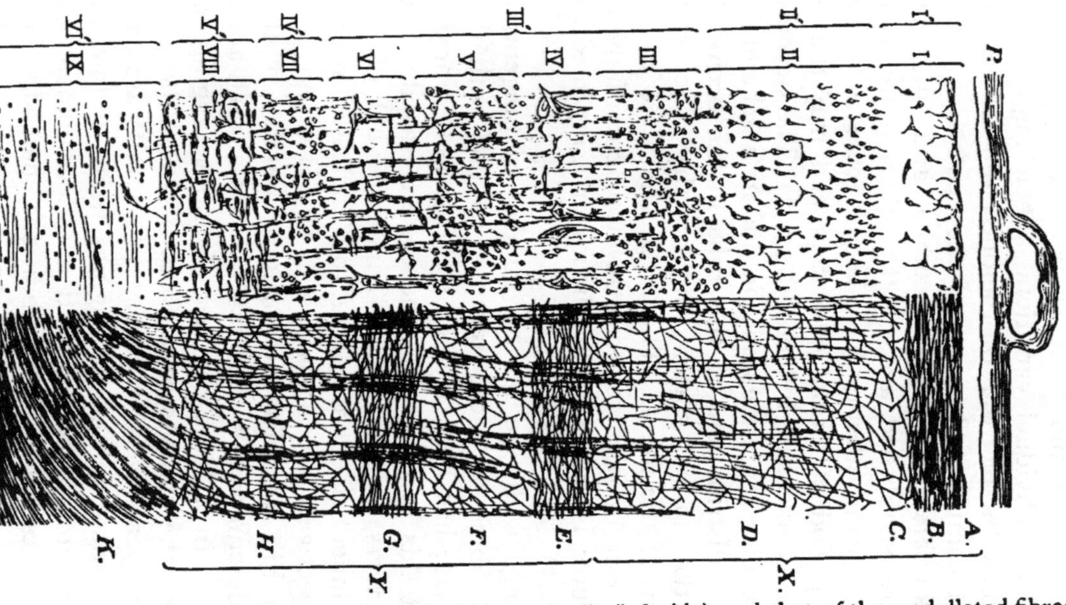

FIG. 47.—To show the arrangement of the layers of cells (left side), and that of the medullated fibres (right side) in the adult human cortex, occipital lobe. (Schematic, Meynert and Obersteiner.) *P*, the layer of pia with a blood-vessel; I–IX, the layers cells as found in the cortex of the sensory regions. (Meynert.) I'-VI', the layers as they would occur in the cortex of the motor regions. The medullated fibres are distributed in the following layers, *A–H*. *X* includes the outer group; *Y*, the inner group of tangential fibres.

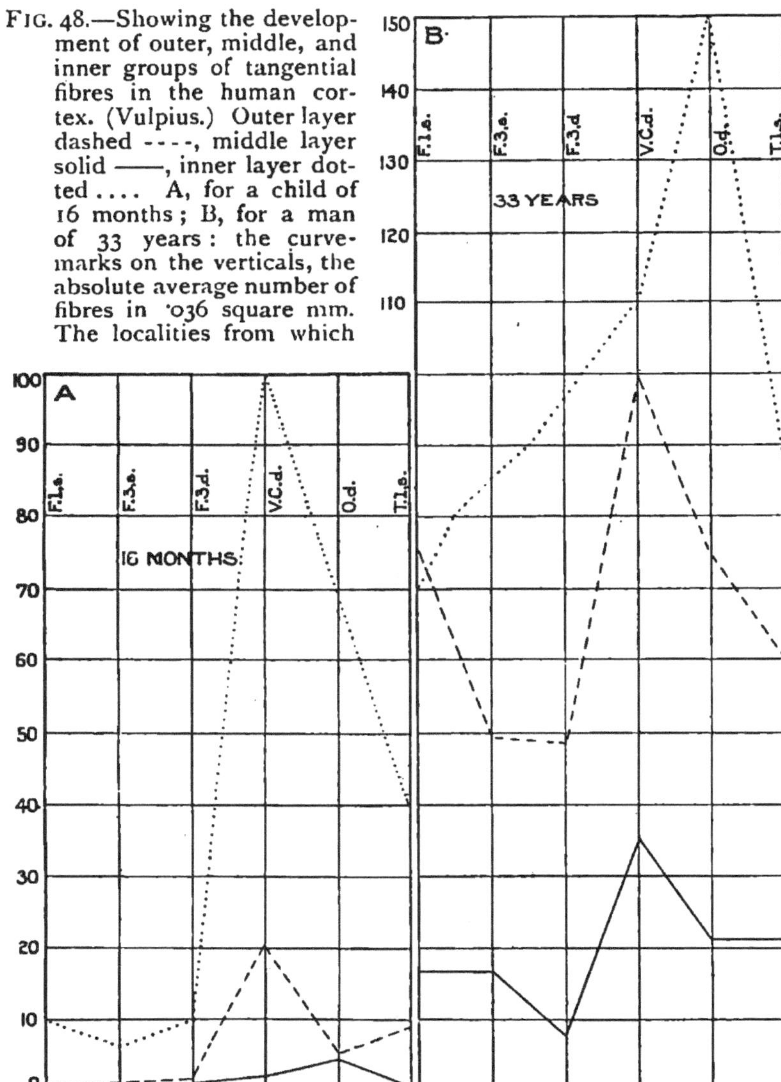

FIG. 48.—Showing the development of outer, middle, and inner groups of tangential fibres in the human cortex. (Vulpius.) Outer layer dashed ----, middle layer solid ——, inner layer dotted .... A, for a child of 16 months; B, for a man of 33 years: the curve-marks on the verticals, the absolute average number of fibres in ·036 square mm. The localities from which the samples were taken are indicated as follows: F.1.s., first frontal gyrus, left; F.3.s., third frontal gyrus, left; F.3.d., ditto, right; V.c.d., anterior central gyrus, right; O.d., occipital lobe, right; T.1.s., first temporal gyrus, left.

layers has increased, and this too despite the fact that the individual fibres have grown larger. At the first age the number is greatest in the anterior central gyrus and the parts behind it. The increase is greatest in the outer layer, next in the middle, and least in the inner one, so that, according to this, the increase in the number of fibres progresses from the deepest layer towards the surface. When, on the other hand, the development at each locality is followed through a number of years, a different, though concordant, series of changes is observed, and these are shown in the three tables given below.

Chart C, F.3s., is a region late to develop, and the growth there is even and regular. Chart D, V. C.d., shows a region well developed at an early age, and since, therefore, the period for this is probably variable, it is not surprising that the curve is uneven ; while Chart E, T.1s., develops more slowly and is again regular. The proportionate increase in the several layers is greatest for those that begin earliest. In the motor areas (F.3s. and V.C.d.) there is a decrease in the number of fibres after the thirty-third year, this being most marked in V.C.d. In T.1.s. this diminution is delayed.

In the new-born child, Vulpius found no medullated fibres in the cortex. In the white substance of the anterior central gyrus alone some medullated fibres were found. The first tangential fibres appear in the inner and outer layers of a child at four and a half months and in the middle layer at eight months.

Along this same line are the observations of Kaes,[1] who directed attention principally to the layers II and and III, Fig. 47 (the middle layer of Vulpius), and aimed to determine how late in life increase in the medullated fibres could be here followed. His method was to

[1] Kaes, *Archiv. f. Psychiat. u. Nervenkrank*, Bd. xxv. 1894.

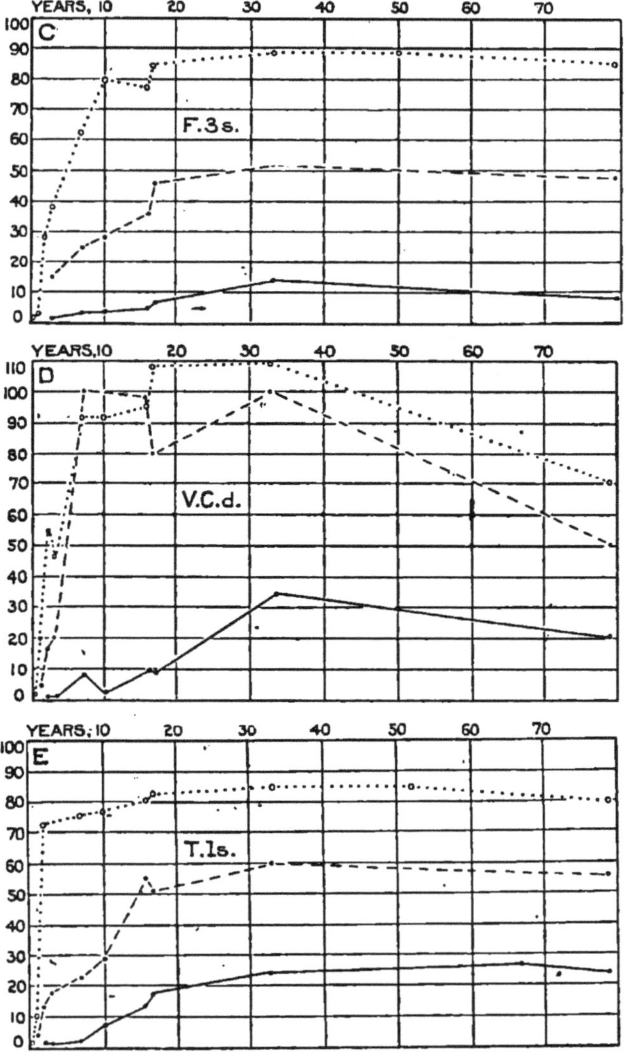

FIG. 49.—Charts C, D, E. Showing the increase in the medullated fibres in the three layers during the growing period, and the decrease in advanced age. Birth, to 79 years (modified from Vulpius). C, third frontal gyrus, left side; D, anterior central gyrus, right side; E, first temporal gyrus, left side. - - - -, outer layer; ——, middle layer; . . . ., inner layer.

measure the thickness of the layers in question, and some differences between his results and those of Vulpius probably depend on the fact that owing to condensation, the number of fibres in a limited area does not always run parallel with the absolute thickness of the layer formed by them.

The medullation is greater in brains of individuals between 38–45 years than in that of a youth of eighteen; so that somewhere in this interval further growth occurs, and from Vulpius we see that it is continued up to at least the thirty-third year. In stained sections in which the layers appear grey, yellow-grey with dark lines, or yellow, according to the abundance of fibres, he found the following proportion of the cortex coloured as indicated in Table 55.

TABLE 55.—SHOWING THE PROPORTIONAL AREA OF THE CORTEX, HAVING THE COLOURS YELLOW, YELLOW-GREY, AND GREY AFTER TREATMENT WITH WOLTER'S STAIN. THE GREYER THE CORTEX THE MORE ABUNDANT THE MEDULLATED FIBRES. (*Kaes.*)

AGE.	PERCENTAGE OF AREA.		
	Yellow.	Yellow-grey with dark lines.	Grey.
Youth, 18 years, right hemisphere...	63	17	20
" 18 " left " ...	59	20	21
Man, 38 " right " ...	37	38	25

This shows the medullated fibres giving the grey tint to be slightly more abundant in the left than in the right hemisphere of the youth of eighteen years, but most abundant in the man of thirty-eight years. The cortex covering the insula was found to be least well developed as regards the layers in question. The fibres were a trifle more abundant in the occipital than the frontal portions. The cortex of the dorsal surface was most developed, that of the ventral least, and the mesal surface was intermediate between these two. The dorsal surface reaches full development first, and the other surfaces follow in the order named. The development is often very local, and there is by no means a uniform character common to large areas.

TABLE 56.—GIVING THE THICKNESS OF THE CORTEX AND ITS SEVERAL LAYERS IN MILLIMETERS. (*Kaes.*)

The letters refer to Fig. 47. The most striking points in the table are the influence of locality and age on the thickness of these layers.

AVERAGE THICKNESS OF THE ENTIRE CORTEX AND ITS LAYERS, IN MILLIMETERS.

	Dorsal Surface.		Mesal Surface.		Ventral Surface.	
Age.	18	38	18	38	18	38
Total width at summit of gyri	3·63	4·01	3·57	3·86	3·43	3·88
,, ,, side of sulci	3·18	3·0	2·53	2·97	2·74	2·94
,, ,, bottom of sulci...	3·05	3·04	2·6	2·86	2·43	2·6
Width of layer containing radial fibres (D-K)	2·91	3·0	2·63	2·8	2·37	3·0
Zonal layer at summit of gyri (B)	0·248	0·24	0·27	0·2	0·239	0·261
,, ,, bottom of sulci	0·72	0·73	0·7	0·72	0·64	0·53
Cell free layer (C)...	0·71	0·61	0·54	0·54	0·7	0·57
I, II and III layer (X)	1·51	1·9	1·6	1·76	1·7	1·7
Baillarger's stripe, outer (G)	0·39	0·345	0·85	0·67	0·8	0·62
,, ,, inner (E)	0·42	0·45	0·33	0·45	0·4	0·35
Outer association layer at summit of gyri (Y)	2·1	2.26	1·9	2·03	1·8	2·17

The specimens showed an increase in the II and III layer by the addition of new fibres and a compacting of the zonal fibres, together with those forming the layers of Baillarger (G), outer or (E) inner, a compression which in the layer first named may possibly be associated with the loss of fibres. These facts are summarised in the accompanying table (56).

There can be little doubt that changes similar to these affect the other portions of the encephalon, although it is not possible at the moment to bring forward direct proof of this. It is evident from these facts that the organisation and therefore the details of the architecture of the central system are continually being modified through life. Somewhere beyond maturity retrogressive changes begin, but the few facts which exist on this phase of growth may be conveniently deferred until the discussion of old age.

CHAPTER XIII.

LOCALISATION OF FUNCTION.

Connections of the central system—Segmental arrangement—Localisation—Multiple innervation—Crossed connections—Relative importance of cerebral centres—Difficulty in the way of a general interpretation—Pathway of fibres—Cortical areas—Motor areas—Arrangement in man—Character of movements—Nature of cortical control—Latent areas—Modification of cortical discharge—Refinement of control—Afferent impulses affecting development—Pyramidal tracts—Sensory areas—Multiple representation—Crossed control—Secondary pathways—Problem of the "unused" hemisphere.

IF we picture the entire nervous system of man composed of a central axis from which nerves pass to all the portions of the body, we shall have the basis from which to start the present inquiry (see Fig. 50).

The roots here represented as arising from the central axis caudad to the pons, contain fibres which transmit not only afferent, but efferent impulses. The fibres which give sensibility to any part, enter the central system near the point from which emerge those that control its movements. As might be expected, therefore, the parts to be controlled are most directly under the influence of the afferent stimuli which arise in their immediate neighbourhood. This arrangement recalls the segmental structure of the central system, whereby both the sensory and motor nerves belonging to a given segment of the body are localised in a corresponding segment of the central system. It was an easy inference

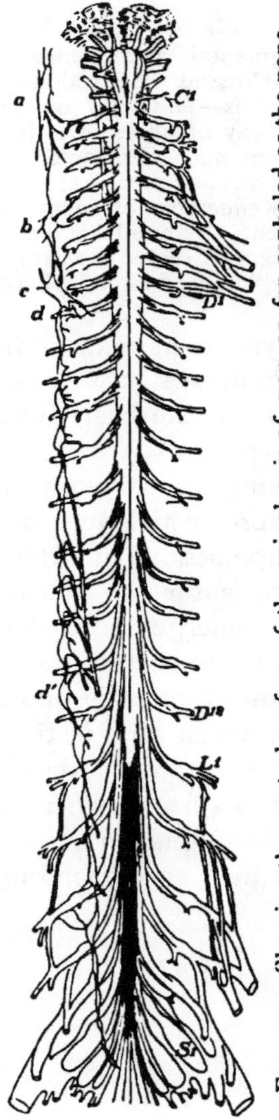

FIG. 50.—Showing the ventral surface of the spinal axis of man, as far cephalad as the pons. The spinal nerves appear on both sides, and on the left the sympathetic ganglia are still in connection with them. (Allen Thomson in Quain's *Anatomy*.) C^1, first cervical root; D^1, first thoracic root; D^{12}, twelfth thoracic root; L^1, first lumbar root; S^1, first sacral root; *a*, *b*, *c*, superior, middle, and inferior cervical sympathetic ganglia; *d*, first thoracic; *d'*, eleventh thoracic.

from this arrangement that the portions of the central system with which a segmental pair of nerves was connected, was also the portion in which occurred the principal central adjustments concerned with this pair, but the question remained as to the significance of those portions of the central system which are more elaborated and are not so plainly segmental. It is naturally asked whether these parts, such as the cerebellum, for instance, are exclusively associated with the nerves which join the axis in their immediate neighbourhood, or whether they have wider and more general connections.

We have seen reason to think when examining the phylogenetic modifications which the mammalian nervous system has undergone, that there is a strong tendency to break down the initial segmental arrangement by condensation, in such wise that groups of centres might act in concert and at the same time any one of them be brought under the combined control of several sense organs. In discussing the ideas of localisation which have developed around these observations, it will be desirable to restrict ourselves here mainly to that side of the question which will throw most light upon the architecture of the central nervous structures.

The architecture has already been examined from the anatomical point of view, as well as embryologically; and it has been pointed out that probably the larger divisions of the encephalon, which are of such interest to us, are the homologues of that region in the spinal cord which forms the dorsal horns and contains a large proportion of the central cells. It remains to be determined how pathways for incoming impulses are there related to those for outgoing ones, and it is in this direction that the physiological experiments yield facts which can be utilised.

But, in the first instance, the problem may be dealt with in a purely statistical manner. Belonging to the spinal cord there are thirty-one pairs of nerves by which impulses enter and leave it. There are also enumerated within the cranium twelve pairs of nerves, of which six are predominantly pathways for incoming impulses, and the other six for those outgoing. I say predominantly, because, as a matter of fact, in the optic and olfactory pathways there are also some fibres which transmit outgoing impulses. Such being the formal arrangement by which the central system is connected with the other parts of the body, we may study a given locality in it with a view to obtaining further details.

Some idea of the central connection of the nerve roots can be got from the arrangement of the motor nerves in the spinal cord and from their relations to the muscles of the limbs. The same muscles may be thrown into contraction by stimulating any one of several of the motor roots. If, then, the group of cells giving rise to these fibres is designated as the nucleus for the nerve fibres controlling that muscle, it is found that such a nucleus may extend through several spinal segments, and that the muscles used in the fine adjustments at the extremity of the limb, those of the hand, for instance, have in the cord extensive nuclei. It appears that by thus drawing into a long column a group of cells which gives rise to the controlling fibres, the muscle is not only put under the direction of a large number of efferent cells, but that since these cells are located at various levels, special groups of them are directly accessible to different incoming impulses; and thus the co-ordinated responses of the muscles are refined. What has been said strictly applies only to the cells giving rise to the efferent fibres. Concerning the distribution of the afferent fibres, there is much less exact information. On this side, however, two relations are well established: first, that by way of the afferent fibres, impulses may directly reach the nuclei of the efferent nerves; and second, that also by way of them, impulses may reach these cells *viâ* the cerebral hemispheres, usually the hemisphere of the opposite side. The accompanying figure illustrates this relation.

The principal path followed by these impulses on the way to the higher centres is either along the dorsal columns of the cord, on the same side as far as the lower end of the bulb, where the impulses, having passed to a new set of cells, cross to the opposite side of the axis and terminate in the corresponding hemisphere; or across the median plane to the lateral column, and so cephalad to the corresponding hemisphere. In this crude manner the impulses may be pictured as distributing themselves within the central system by means of the incoming fibres and their collateral branches, aided by the central cells. The expansion of the nucleus of a motor nerve through several segments is, by comparison, centralisation in a high degree when contrasted with the manner in which a sensory pathway thus extends itself over a large part of the central system. It is well to bear in mind that the cranial nerves, so far as they mediate afferent impulses, have, like the sensory nerves of the spinal cord, primary centres at the points where they terminate in the axis, and that from those primary centres they establish secondary connections with the cerebral hemispheres. They resemble the sensory nerves of the spinal cord also in the fact that the

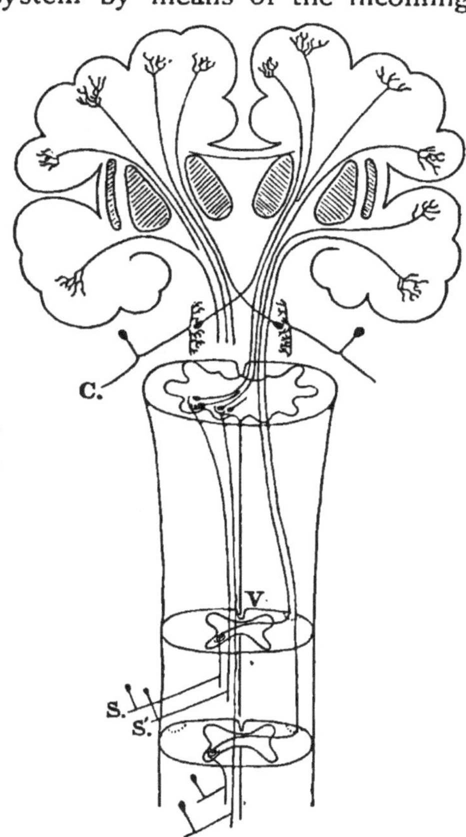

FIG. 51.—Schema showing the pathway of the sensory impulses. On the left side, S, S', represent afferent spinal nerve fibres; C, an afferent cranial nerve fibre. This fibre in each case terminates near a central cell, the neuron of which crosses the middle line, and ends in the opposite hemisphere. (Van Gehuchten.)

crossed relations which they finally attain with the cerebral hemispheres are due to the interpolation of fibres arising from the central cells.

The study of the lower vertebrates after injury to the

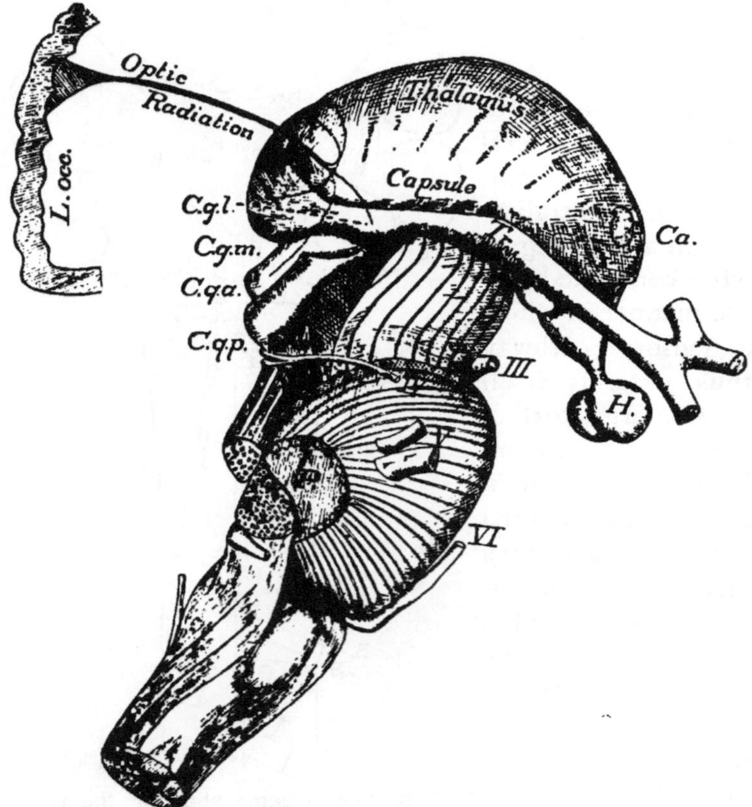

FIG. 52.—Showing in man the connection of the primary centres for vision with the occipital cortex by means of the optic radiation. (Edinger.) *Ca.*, anterior commissure; *C.g.l.*, lateral geniculate body; *C.g.m.*, mesial geniculate body; *C.g.a.*, anterior quadrigeminum; *C.g.p.*, posterior quadrigeminum; *H.*, hypophysis; *L.occ.*, occipital lobe; *P.*, pons; *Tr.*, optic tract; *III.*, oculo-motor nerve; *IV.*, trochlear nerve; *V.*, trigeminal nerve; *VI.*, abducent nerve.

different divisions of the central system shows that in

those forms in which cephalisation is but little advanced the primary centres of the cranial nerves when alone present may assume a guiding control over the remainder of the system. It thus happens that a frog after loss of the cerebral hemispheres can still direct its jumping movements so as to avoid a visible obstacle in its path; in other words, impressions reach the central system of such a frog through its eyes, and these impressions influence the reactions of the muscles of hind legs despite the absence of the hemispheres. In man, on the other hand, the parts of the brain corresponding to the optic lobes of the frog do not represent a locality in which such connections are established, so that in him the hemispheres alone do the work which in the less specialised form may be performed by the lower centres. In this connection we naturally inquire how the cerebral hemispheres may have acquired in the higher vertebrates capabilities which belong to them in a less and less degree as we descend from man through the zoological scale. In the higher forms it appears that the incoming impulses, instead of passing over in the primary centres to cells which discharge downwards, pass to a group of afferent central cells which carry impulses to the cortex, that with the organisation of this second pathway the first becomes less passable, and thus the function is transferred, though the causes determining the growth of the central cells on which the change depends are still obscure.

One difficulty felt by all who have touched this problem of localisation is due to the very large portion of the encephalon, especially in the cerebrum, to which our observations and conclusions have not yet been applied. At first explanation is incommoded by the existence of these enigmatic parts, but only too readily do we neglect what cannot be easily explained, and by so doing really hamper further advance. The partial character of the explanations which are to be given should therefore be kept in mind.

The tracts connecting the hemispheres with the spinal cord, and bearing incoming impulses, run mainly through the dorsal portions of the central axis. On

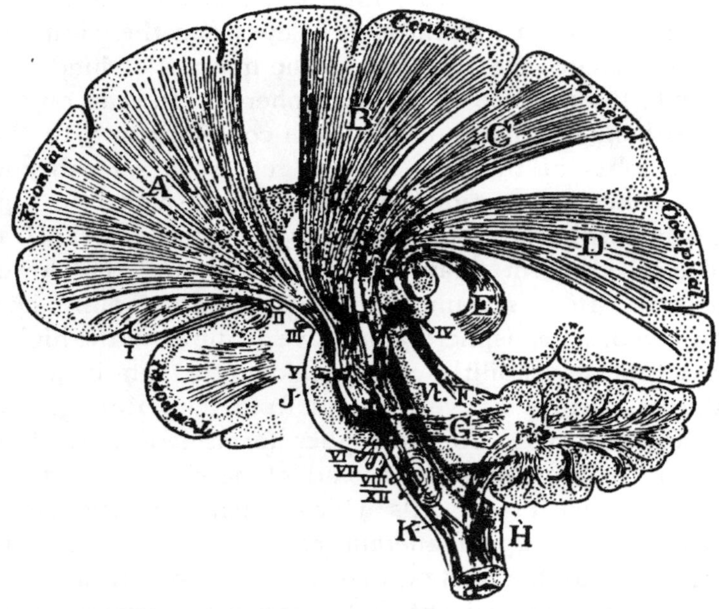

FIG. 53.—Schema of the projection fibres within the brain (Starr.) Lateral view of the internal capsule. A, tract from the frontal gyri to the pons nuclei, and so to the cerebellum; B, motor tract; C, sensory tract for touch (separated from B for the sake of clearness in the schema); D, Visual tract; E, auditory tract; F, G, H, superior, middle, and inferior cerebellar peduncles; J, fibres from auditory nucleus, the inferior quadrigeminal body; K, motor decussation in the bulb; Vt, fourth ventricle. The numerals refer to the cranial nerves. The sensory radiations are seen to be massed towards the occipital end of the hemisphere.

passing from the brain stem into the hemispheres these tracts, in company with other fibres bearing incoming impulses from the cranial nerves, enter the locality which is called the internal capsule.

In this region it is the occipital end of the capsule which is most completely occupied by the afferent

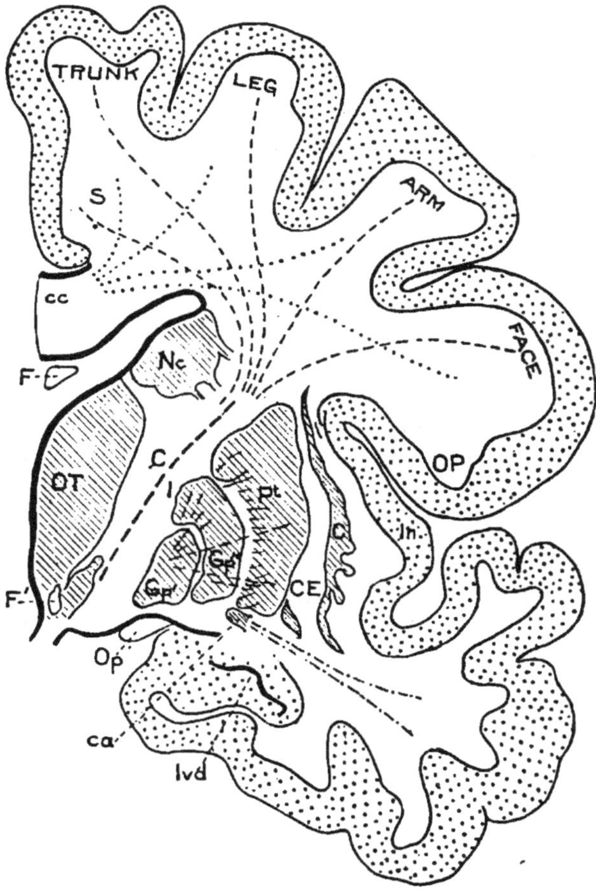

FIG. 54.—A frontal section of a human hemisphere, showing the internal capsule. C, a mass of fibres passing between Nc and OT on the one hand, and Pt., Gp', and Gp" on the other. Natural size (Sherrington, Foster's *Physiology*) ; NS, Caudate nucleus ; OT, optic thalamus ; Gp", Gp', two parts of the globus pallidus ; Pt, putamen.

fibres (see Fig. 61). The relations of the fibre bundles in the capsule are sometimes so described as to leave the impression that this is the sole portion of the capsule in which such afferent fibres occur. More probably, however, they are distributed through its entire extent because the method of distribution in the cortex, like that for the spinal cord, is such that the fibres bringing afferent impulses terminate in the neighbourhood of cell-bodies whose neurons control the centres for the related muscles, and further, because the general grouping of the fibres in the internal capsule itself is in the main similar to the grouping in the cortex. The distribution of the incoming fibres is consequently such that a large portion of the cortex must receive them, and, judging from its reactions to stimuli, an equally large region must contain cells giving rise to outgoing impulses. The same reactions which lead to the conclusion that the incoming impulses are brought to the surface of the cerebrum lead also to the conclusion that at this point they pass from one set of nerve cells to another, and that this second set discharges towards the lower lying centres. Thus there is indicated an arrangement by which at the cortex the direction of the nervous impulse is exactly reversed. Such changes of direction may occur in any centre, but the peculiarity of the cortex from the standpoint of physiology consists in the extension over a broad surface of these turning-points for the nerve impulses, an extension which particularly adapts the cortex for experimental study. Granting these points, it is not surprising to find that in a certain sense the entire cortex, so far as it responds directly, is motor, so that the stimulation of it at different points will give rise to muscular contractions. This close association of the two groups of elements brings it to pass also that in the case of the regions

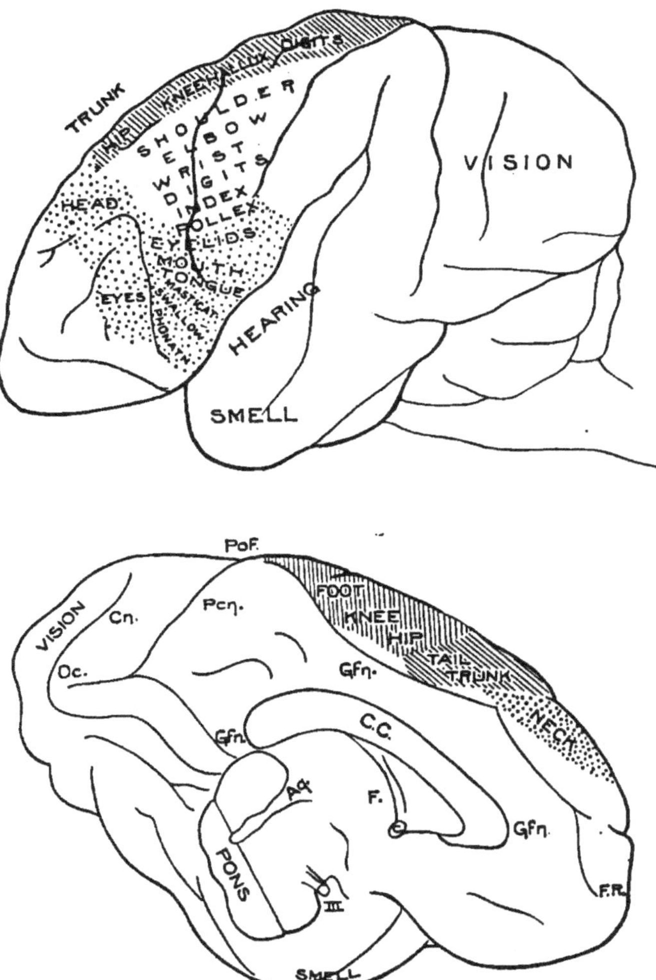

FIG. 55.—Brain of the macaque monkey, showing the sensory and motor areas. In the sensory region the name of the sensation is over the locality most closely associated with the corresponding sense organ. In the motor region, the name of the part is written over the portion of the cortex which controls it. The uppermost figure gives a lateral view of the hemisphere and the lowest a mesal view. (Beevor and Horsley, from Foster's *Physiology*.)

devoted to special senses, like sight, hearing, and smell, the immediate motor reactions which arise from the stimulation of these areas occur in the muscles controlling the corresponding external sense organs. On the other hand, the mass of muscles which has for its principal sense organ the skin, is by comparison with the muscles of the external sense organs very large, and the locality from which the reaction of the former can be obtained is so elaborated that this region has been called motor, to distinguish it by contrast from the other regions designated as sensory. Such designations, therefore, as motor or sensory are misleading, and although the sensory side in one locality may be most highly developed, and the motor in another, yet in all of them are to be found pathways for both incoming and outgoing impulses.

Before discussing the cortical areas in detail, these great regions, as they appear in the brains of the monkey (and probably of man), should be described. The monkey is chosen for comparison because the most important experiments which have been made in this direction have been upon a species of macaque (*Macacus sinicus*), and though in a general way the relations in this animal are similar to those in other monkeys and in man, yet, as there are variations in detail, it is well to remember the species of animal which has been employed. Fig. 55 shows the relation of the motor and sensory centres in this monkey as determined by Horsley.

As a matter of convenience regions may be termed motor or sensory, if only it is remembered that by these terms we indicate the more evident rather than the exclusive character there found. There are two peculiarities in the distribution of the motor areas in the monkey which are worthy of remark. In the first place, a large proportion of the convex surface of a hemisphere is devoted to the control of the muscles whose sense organ is the skin, and in the second place, this area extends on to the mesal surface of the hemispheres. If we compare with this the arrangement in the brain of the orang, we find the motor areas in question more concentrated about the central fissure, or fissure of Rolando (the long fissure on either side of which they are seen to lie in Figs. 55 and 56).

Comparing the arrangement in the monkey with that in the orang, it is found that the proportion of the cerebral cortex devoted to these areas is smaller in the orang, and that within the larger areas, at least, there are spots the stimulation of which does not yield any reaction. The experimental determination of these areas made on the brain of man are of course few in number. So far as they go, they indicate that in this respect the human brain shows an increase in the concentration and isolation of these areas, and of the centres within them. There is certainly no more important feature of the whole subject of localisation than this concentration in man and the higher apes, but it may be for the moment disregarded while the characters which these areas have in common are first described.

As might be expected, the homologous portions of the cortex in man and the apes are occupied by similar centres. The motor regions, for example, are grouped in all cases about the central fissure. The area for vision is in the occipital lobe, and that for hearing in the temporal. Not only does this general distribution show similarities with that in other mammals, but the similarities extend to many details.

The areas for the different portions of the body are so distributed in the motor cortex that those controlling the muscles of the head and neck lie most anteriorly, while those controlling the muscles of the leg and foot lie most posteriorly, with the areas for the upper arm and trunk between them. The actual relation of these

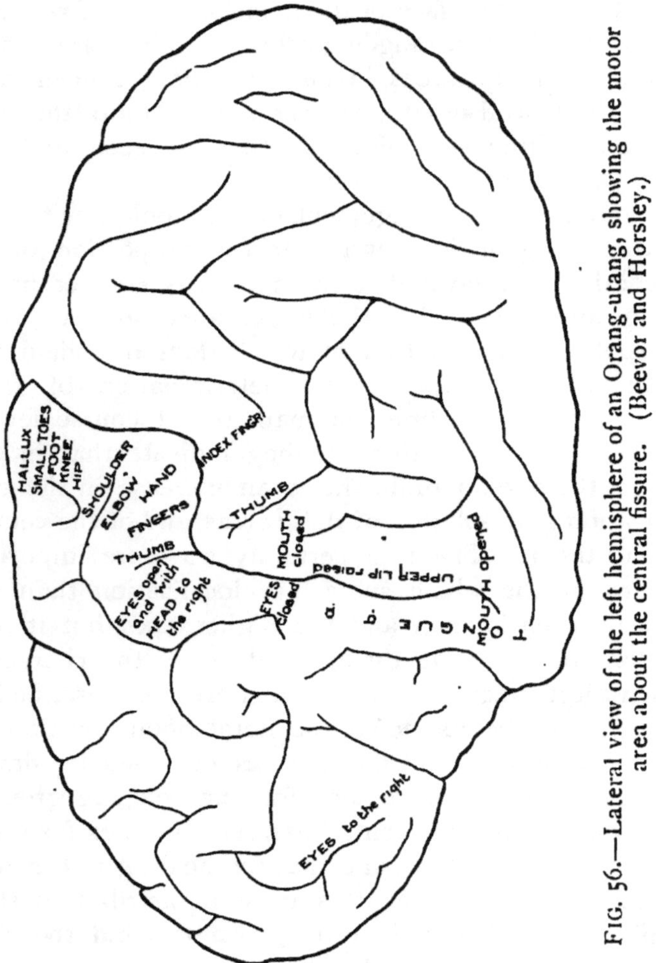

FIG. 56.—Lateral view of the left hemisphere of an Orang-utang, showing the motor area about the central fissure. (Beevor and Horsley.)

areas is certainly not perfectly schematic, but there is sufficient order among them to warrant the statement that the nerves bringing impulses from the sensory areas widely separated on the surface of the body tend to be distributed in widely separated portions of the cortex. This relation between the larger areas is repeated in a very remarkable way by the smaller centres composing them. It is there found, as can be seen by consulting the diagram, Fig. 55, that the point of the cortex, the stimulation of which gives rise to movements of the muscles controlling the most proximal joints of a limb, the shoulder, for example, is farthest separated from the portion which controls the most distal joints, those of the fingers, and that the intermediate joints are represented by centres lying between them. In this respect, also, the separation of parts at the periphery of the body is accompanied by a corresponding separation in the cortex.

The division of the motor areas, according to the segment of the limb in which the movement is initiated, is, however, but one way of dividing them. An extremely suggestive subdivision is that made according to the general character of the movements following its excitation, these movements being classified as those of extension, confusion, or flexion. In the arm area it is found to be the upper and frontal part which most readily gives rise to movements of extension, while the lower and occipital portion gives rise to those of flexion, an area for confused movements lying between them, and this separation of opposite movements probably holds for those of the leg area also.

The more carefully the relations of the cortex to the lower centres are studied the plainer it becomes that the impulses which pass down from it act on them as but one of several groups of stimuli, comparable with those arriving directly from the periphery, and subject to modifications by them. The cortical centres do not usually control individual muscles, but groups of muscles, the contraction of which results in co-ordinated movements, and thus the nature of the reaction taking place in consequence of the cortical discharge is modified first by alterations of the initial stimulus during its passage through the cortex, and second, by the conditions of the lower centre as determined by impulses arriving from other sources.

Granting the relation here suggested, we are prepared

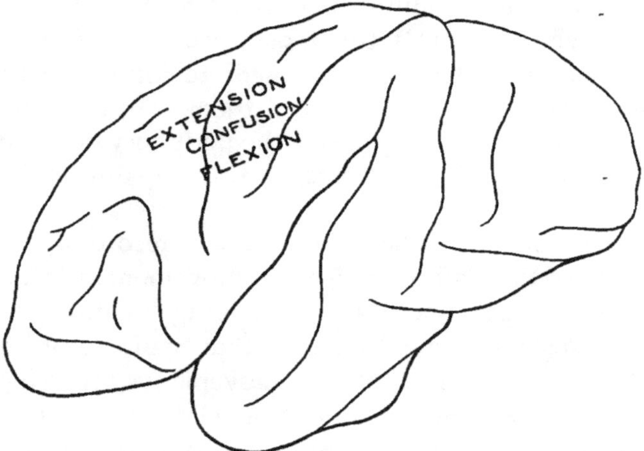

FIG. 57.—Showing the localisation of movements of different characters in the arm area. (After Horsley.)

to find that the cortical representations of different muscle groups is not at all dependent on their mass, but is related rather to refinement and complexity of the movements which these muscles effect. Accordingly the massive muscles of the trunk and leg are but poorly represented, while, on the other hand, those of the face and arm have much larger areas devoted to them. And further, in the arm area it is to the thumb and index finger, in the head area to the tongue, lips, and muscles of phonation, to which the greatest extension is given. The exact adjustment of the refined movements represented by these groups of muscles demands many associations, and it is apparently as an anatomical device for furnishing such associations that their extensive representation is significant. Experimentally it is possible to partially innervate a muscle, throwing into contraction only a portion of its fibres, and in making delicate and finely adjusted voluntary contractions it is probably necessary to innervate the different groups of muscle fibres in varying proportions. The mechanism for the tongue may be taken as an illustration. To the tongue run many fibres from the spinal centres, and since there are many they must come from different portions of the motor nucleus. The tongue is moved whenever the appropriate area in the cortex is stimulated, but the contraction is slightly varied according to the portion of the area aroused, and under normal conditions the incoming impulse associated with one form of movement will act through one group of efferent cells, while another movement will depend on the discharge of a slightly different group, thus giving an anatomical basis for differences in the resultant contractions.

It is to be noted in this connection that the movements of the body as a whole do not belong to the group requiring fine adjustment.

As already stated, experiment shows that in man the special cortical centres are somewhat separated from one another, and this separation is due to masses of brain substance which do not give reactions upon ordinary electrical stimulation. Further than this, the proportion of the brain which is thus latent, giving no reaction upon stimulation, seems to be larger in man than in the monkey. Certain differences between man and the macaque monkey in the development of the cortical centres can be explained by the fact that the macaque monkey makes a greater voluntary use of its legs in climbing. In man this control is not exercised, and therefore there is no corresponding development, yet it is by no means easy to say what the increase in the latent areas may signify. From what is known of the human brain it is justifiable to consider this peculiarity as a mark of superiority. It appears, therefore, that the development of the human cortex is not, as might perhaps have been anticipated from the study of the lower mammals, dependent on an increasing localisation, whereby larger and larger portions of the cortex were given over to the efferent control of the different groups of muscles, but to a limitation of the areas from which the efferent impulse can be aroused. Possibly the failure to obtain a response from so much of the human cortex is related to the larger development of the central elements having an associative function, about the several groups of efferent elements, the artificial stimulus being therefore applied to but a fraction of the central elements usually concerned in the discharge of the efferent group. This view has been recently expressed by Flechsig.[1] Moreover, a distinction can fairly be made between muscular reactions like those of facial expression, which are in themselves highly variable, and the habitual movements of the leg which, though not so variable in itself, may yet occur in almost any combination with other movements. It is hardly to be expected that the cortical arrangement would in these two cases be similar.

Thus far attention has been directed to the efferent cortical discharge which would occur in consequence of incoming impulses directly arriving in the immediate neighbourhood of the efferent cells. At the same time, it is recognised that impressions received through the ear can influence, let us say, the movements of the

[1] Flechsig, *Neurolog. Centralblatt*, No. 19, 1894.

leg, and it is therefore argued that there must be connections between these widely separated areas. It would appear that this connection depends upon pathways of association furnished by central cells, and along which the disturbance set up in the auditory area is communicated to the leg area. The relation here cited is but one of many combinations which can occur, and we have to conceive the incoming disturbances as distributing itself in many directions, although the response

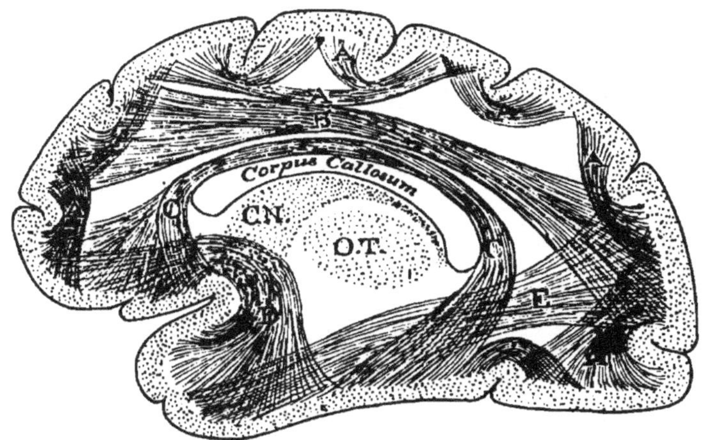

FIG. 58.—Lateral view of a human hemisphere, showing the bundles of association fibres. (Starr.) A, A, between adjacent gyri; B, between frontal and occipital areas; C, between frontal and temporal areas, cingulum; D, between frontal and temporal areas, fasciculus uncinatus; E, between occipital and temporal areas, fasciculus longitudinalis inferior; C.N., caudate nucleus; O.T., optic thalamus.

to it may be limited. Some of the larger pathways which probably furnish such connection are shown in Fig. 58.

In the cortex, therefore, as in the case of the spinal cord, a sensory impulse arriving at one of these given points, may diffuse itself in many directions, and indeed must do so, if it is to bring several motor centres under its influence. The sensory impulse reaching the cortical cells may thus be compared to a complex sound wave striking upon resonators, each one of which picks out that vibration to which it has been attuned and responds to it. Moreover, to push the simile further, the pitch of responsive cells may be altered by the play of other impulses upon them, and thus the analysis at different times is not the same. Refinement in the structure of the cerebral cortex may, therefore, be developed in three ways: first, by the multiplication of the pathways bearing the incoming impulses; second, by rendering more sensitive to slight differences in the stimulation those cells whose function it is to receive these impulses; and third, by increasing the number of the central cells. So far as can be seen at present, the brains of the lower and less intelligent mammals are inferior in all these respects, but are most deficient on the side of the afferent and central elements.

Besides the prime duty of furnishing pathways for the incoming impulses destined to arouse sensations, the afferent elements take part in other changes which are of great importance. On them primarily depends the full and final development of the cortex, for, to some extent, the very growth of the hemispheres hangs on the manner in which they are excited, healthy activity promoting fuller development. It is not necessary to go into details which favour such a view; but much evidence can be obtained from cases in which the course of development has been more or less modified, as the result of interference with the incoming impulses.

As might be supposed, the pathways connecting the efferent cells in the cortex with the lower centres in the central axis, are proportionately more numerous in man than in any other animal, when we measure them by means of the area which they occupy in cross-sections of the bulb or cord, for since the fibres are in the several cases approximately of the same diameter the area of the bundle corresponds to the number of fibres which it contains. The relations are shown in the accompanying figure.

These anatomical relations are inadequate to fully measure the physiological differences, for man surpasses the dog in the adjustment of his limbs to a greater degree than is here indicated, because the total

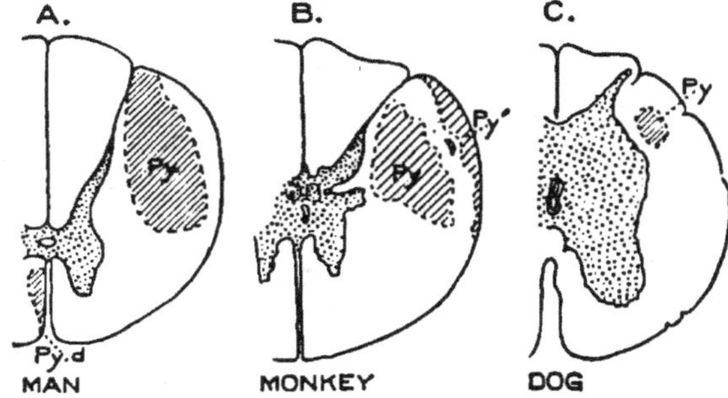

FIG. 59.—Schemas of a cross-section of the spinal cord in the dorsal region in three mammals, the sections being all enlarged to the size of that representing man, which has been magnified 4 diameters. The cross-lined portions indicate the area of the pyramidal tracts. (Sherrington, in Foster's *Physiology*.)

complexity of the reactions must be expressed by the increased complexity of all the structural elements involved in producing it, and not that of any one alone.

Since the fibres of the pyramidal tract take origin in the efferent elements of the cortex, the greater the extent of the area the more abundant should be the fibres coming from it. Thus it is evident that the bundle of fibres coming from the large cortical area for the arm should have a greater cross-section than that coming from the much smaller cortical area for the leg and examination of cross-sections of the cord supports

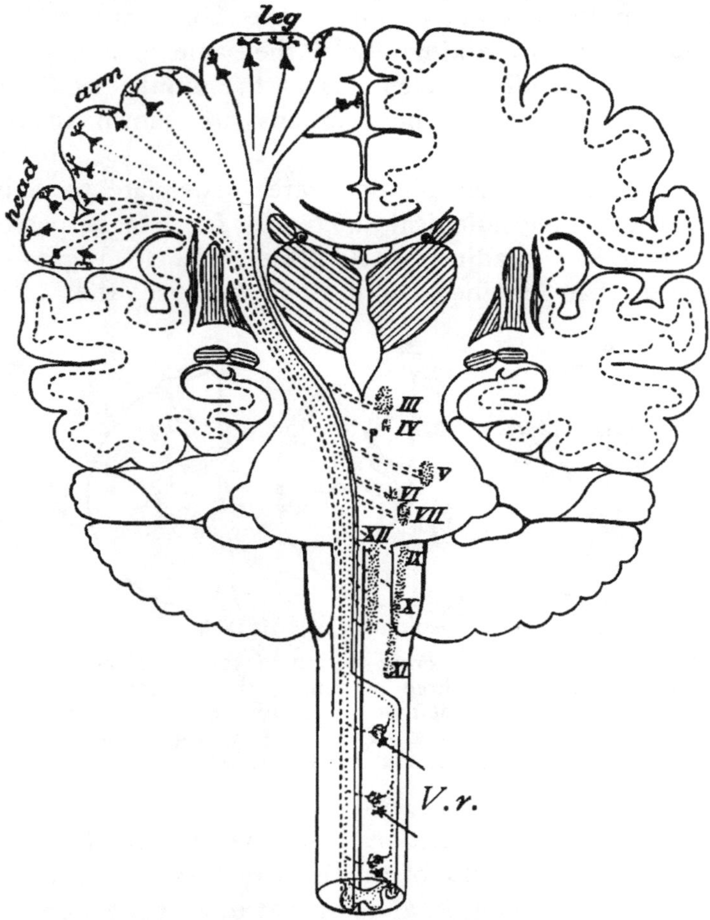

Fig. 60.—Schema showing the pathway of efferent fibres. (Van Gehuchten.) *III–XII*, nuclei of cranial nerves in connection with "head area" of the cortex; *V.r.*, central root of the cervical region in connection with the "arm area" of the cortex.

this inference. These fibres forming the pyramidal tracts pass down towards the spinal cord by way of

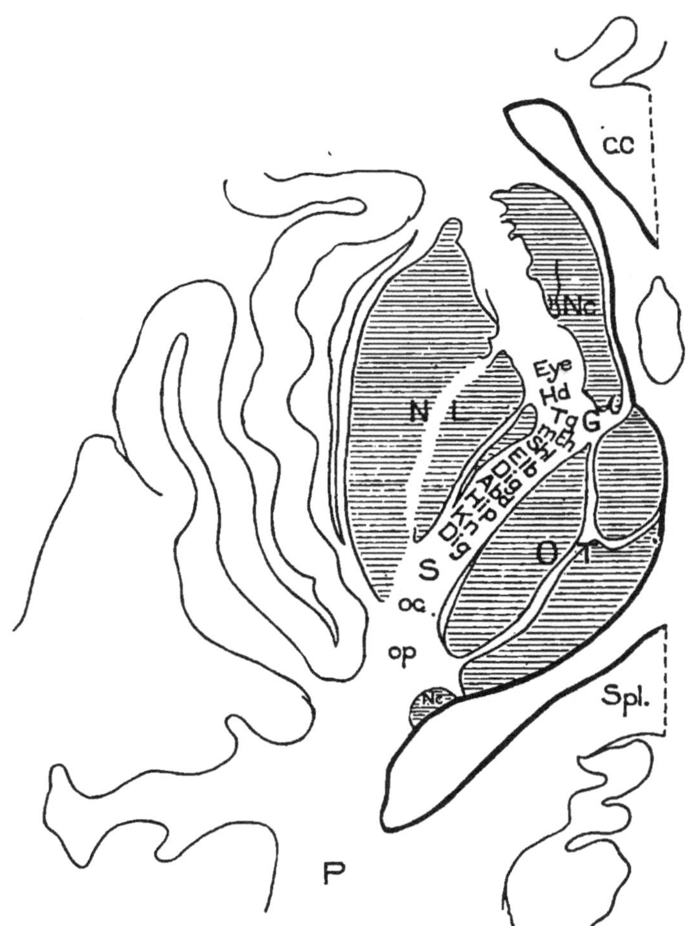

FIG. 61.—A horizontal section of a human hemisphere, showing the internal capsule, a band of fibres lying between Nc and OT on the one hand, and NL on the other. S, placed towards the occipital end, represents the locality where the afferent fibres are most abundant. Slightly enlarged. (Foster's *Physiology*.) CC, callosum ; G, knee of the internal capsule ; Nc, caudate nucleus ; NL, lenticular nucleus ; OT, optic thalamus ; P, occipital end of section ; Spl., splenium. Within the capsule the several bundles of fibres are arranged as follows : fibres for the eye muscles, Eye ; for the head, Hd ; the tongue, Tg ; the mouth, Mth ; shoulder, Shl ; elbow, Elb ; digits, Dig ; abdomen, Abd ; hip, Hip ; knee, Kn ; toes, Dig ; S, the temporo-occipital tract ; oc, fibres to the occipital lobe ; op, optic radiation.

the internal capsule. Their position in the capsule, both with relation to the rest of the brain and to one another, is highly constant, and there is here found a degree of localisation which is just as exact as that in the cortex itself. This arrangement is shown in Fig. 61.

It has been seen that the larger number of incoming fibres cross the middle line at the lower end of the bulb and enter the opposite hemisphere. In a like manner, the fibres arising from this hemisphere and carrying the outgoing impulses cross with one another at nearly the same level, and so innervate the nuclei controlling the muscles of the opposite half of the body.

Just as experiments have shown in the case of the so-called motor areas, that removal of the cortex is accompanied by disturbances of the dermal sensations, so in the sensory areas removal of the cortex affects one or the other of the sense organs named. By this method, combined with pathological observations, we learn that the sense of smell is probably associated with the mesal surface of the temporal lobe, the sense of hearing with the first and second temporal gyri in the middle part of their course, and the sense of sight with the occipital lobe, and especially with that portion of it called the cuneus. The paths by which the stimuli arriving along these different lines may reach and affect any of the efferent cells elsewhere located in the cortex, are probably found in the great systems of association and commissural fibres, already described.

Any sensation can serve to arouse any group of motor centres which are under voluntary control. At the same time certain groups of incoming impulses are usually accompanied by corresponding reactions, or in other words, that there are preferred pathways between the incoming fibres and discharging cells representing those muscular adjustments which are easiest.

Some movements like those for the eyes are represented by more than one point in the cortex, from which they may be aroused by direct stimulation. It appears that the movements of the eyes, which follow stimulation of the frontal region, are due to impulses, which, arising there, pass to the lower centres, from which arise the motor nerves controlling the muscles of the eyes; and in the same way movements of the eyes, which follow stimulation of the occipital region, depend on impulses which pass directly from the point of stimulation to the lower centres. It has been suggested that when the sensory area in the occipital region was stimulated, and movements of the eyes followed, they were due to impulses which passed from the occipital to the frontal region by way of the association fibres, and then through the mediation of the frontal centres found their way to the lower centres. But Schäfer showed that in the monkey's brain the two cortical[1] regions might be completely separated from one another, and yet the reaction followed stimulation of the occipital cortex, thus proving that they were not connected. At the same time, the movements of the leg, for example, can be modified by impressions received through the eyes, and in this case the muscles are in part controlled by a sense organ with which they are not associated at the periphery, and whose cortical area is remote from that of the efferent fibres. In the reactions of which this is a type, as when one steps back before a wave of water, the reaction is probably mediated by association fibres between the sensory, visual, and motor, leg areas. This is very important as a fundamental arrangement, because, while the primary condition puts each group of muscles into relations with the sense organ with which it has the most immediate peripheral associations, this second

[1] Schäfer, *Proc. Roy. Soc.*, London, 1888.

arrangement renders it possible for every sense organ to control the centre for each group of muscles, in so far as it is connected with it by paths of associating fibres, and renders possible even the cross associations between the different sensory areas, which are indicated by coloured hearing and the like.

Although anatomically the evidence is good that the nerve tracts which put one hemisphere in connection with one half of the spinal cord, cross the middle line at the lower end of the bulb, yet physiological experiments, in which one hemisphere of the cerebrum has been removed from a dog, show that both halves of the body may come under the voluntary control of the remaining half of the cerebrum. This relation is not readily explained. One difficulty resides in the fact that we can by no means be sure that the manner in which the impulses pass in the system after the operation is the only one by which they pass under normal conditions, for the central system gives evidence of possessing supplementary pathways. Keeping this in mind, it appears that the pathway by which one hemisphere can control both halves of the body is dependent on a cross connection between the appropriate lower centres in the spinal cord. This cross connection is such in those cases in which the muscles act in unison, as in the muscles of mastication or phonation, that injury to one hemisphere causes but little disturbance in the performance of these muscles, the voluntary control of the centres on both sides being easily maintained from either cerebral hemisphere. On the other hand, where symmetrical groups of muscles do not customarily react together, the control of them by one hemisphere is but slowly attained, even in the dog.

It would appear to be most easily accomplished in young and half-grown animals—a fact which can be interpreted to mean either that before maturity there exist pathways which later cease to be permeable, or, the lesion occurring before maturity, that under the new and abnormal condition fresh connections are established by compensatory growth, the capacity for which growth is later lost. In man, of course, the paralysis of a limb caused by injury to the adult cortex is permanent, and there is no suggestion that the other hemisphere acquires a subsequent control of both halves of the cord, but a similar lesion may be far less damaging when it occurs in a child.

This leads to the question of the relative importance of the two cerebral hemispheres in man. It is probable, from all that can be ascertained, that in a thoroughly ambidextrous individual the two hemispheres more nearly correspond in their functions than they do in the one-handed individuals, as represented by the majority of the community. It is certain, however, that while in the strongly right-handed persons it is the left hemisphere which is mainly concerned, the reverse is the case in those left-handed. For example, in the average right-handed person the entire series of disturbances which are grouped under the term aphasia, occur as the result of injuries to the left hemisphere. Seeking to explain this, we recognise that the sense organs connected with injured hemispheres are most naturally associated with the motor centres of the same side ; but even if this is granted, the manner in which this peculiarity was acquired, and the duties of the other hemisphere, remain still unexplained. Though in children injury to one hemisphere may be compensated by the development of the other, in the adult such is not the case. Normally, too, the hemispheres attain nearly the same weight, and there is no evidence that the left hemisphere is persistently the heavier in a right-handed person. The reasons for these relations are, therefore, not evident in the present way of regarding the nervous system, according to which growth and increase in size are associated with activity. The anatomical arrangement, which was originally responsible for this one-sidedness, has still to be investigated, in order to determine whether the better development of the afferent or efferent structures control the matter in the first instance, and to discover, if possible, how far the physiological processes in the neglected hemisphere may be duplicates of those in the one preferred. At the moment, however, it is not possible to do more than state the difficulty. Yet it is possible that since the relations between the outgrowths of one cell element and those of another must be both very exact and very close, a cessation of the growth processes just this side of the best degree of approximation between the two would produce a defect in organisation, not accompanied by differences in weight, since they might easily be masked, in the sense that a slightly better development of myeline might occur on the neglected side. Such questions as are suggested by aphasia, the analysis of which has contributed so much to our knowledge of brain processes, can be discussed more conveniently in the chapter on education, and will be there treated after the activity, fatigue, and old age of the central system have been reviewed.

CHAPTER XIV.

PHYSIOLOGICAL CHANGES IN THE CENTRAL SYSTEM.

Nerve cell a source of energy—The nerve impulse—Direction of impulse—Nerve impulse aroused at the end of branches—Strength, form, and rate of impulse—Arousal—Pathway—Continuity of stimulation—Diffusion—Reinforcement—Diffusion illustrated—Knee-kick—Background of sensation.

THE nerve element is a source of energy. Composing the nerve cell are substances in chemical equilibrium so unstable that stimuli easily excite them to further change. As a consequence they assume a more stable chemical form, and at the same time part with some of their energy. Various similes have been used to illustrate this event. The irritable cell is often pictured as a powder magazine to which a small spark, the stimulus, may be applied, with the result of causing an explosion that liberates a quantity of force many times greater than that represented by the spark. The powder magazine, however, explodes with the same force whether it be ignited by a match or by a torch, and in such a case, therefore, no proportion obtains between the magnitude of the stimulus and the magnitude of the explosion. For this reason the simile does not fully express the facts, since in the nerve cell the larger stimulus is followed by the larger explosion. A truer picture may be obtained by comparing these events with the result of rolling a ball against a group of pins.

In this case the resulting breakdown depends in part upon the force developed by the falling pins, but the amount of this is in turn dependent on the force with which the ball—corresponding to the initial stimulus—strikes them. The actual disproportion which exists between the stimulus and the reaction is less emphasised by this illustration than by that of the powder magazine, but it possesses the advantage of indicating how the energy of the final disturbance depends in part on that of the stimulus. In view of this relation the possible variations in the strength of the impulse which may cause these changes become important.

The change set up in the nerve by exciting it is designated the nerve impulse. Of its intimate nature nothing is known. The impulse has been investigated, however, by the aid of the electrical changes in the nerve which accompany it like a shadow, and make the study of it possible. All portions of the nervous system are irritable, which is equivalent to saying that the stimulation of any portion can give rise to nerve impulses; and it has been further found that when artificially aroused in nerve fibres the impulses travel in both directions from the point of stimulation.

Attention is here claimed for the important fact that the nerve impulse is normally not aroused in the cell-body, but at the ends of the cell outgrowths, so that however much these outgrowths differ from the cell giving origin to them, irritability and the capacity for transmitting the nerve impulse must be granted to them. There can be little doubt that in a given nerve element the impulses have a predominant direction, but this arrangement does not exclude the possibility of other impulses whose direction is the reverse (see Fig. 62).

Directly bearing on this suggestion are some ex-

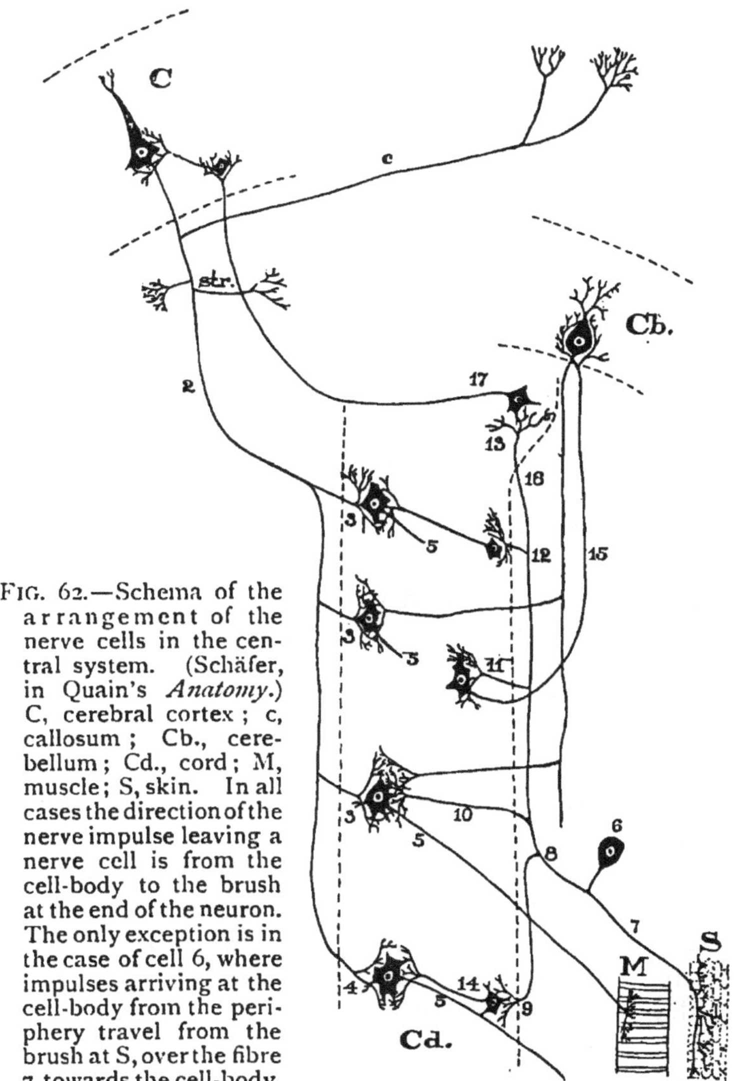

FIG. 62.—Schema of the arrangement of the nerve cells in the central system. (Schäfer, in Quain's *Anatomy*.) C, cerebral cortex; c, callosum; Cb., cerebellum; Cd., cord; M, muscle; S, skin. In all cases the direction of the nerve impulse leaving a nerve cell is from the cell-body to the brush at the end of the neuron. The only exception is in the case of cell 6, where impulses arriving at the cell-body from the periphery travel from the brush at S, over the fibre 7, towards the cell-body.

periments by Gotch and Horsley, which show that the nerve impulses may pass out of the spinal cord by way of the dorsal roots, although usually they are pictured only as passing in.[1] On the other hand, these observers found that if a ventral root were stimulated the nervous impulse could not be detected as passing in any of the columns of the cord; apparently it was blocked in the cell-bodies of these ventral elements. This fact does not prove, however, that such an impulse is without influence in modifying both the reactions and the nutrition of the cell-body in which it seems to disappear.

These beautiful researches of Gotch and Horsley have shown that the strength of a nerve impulse as measured in the columns of the spinal cord is much diminished when it emerges through a ventral root. There may be several causes for this, but the fact of interest is the smallness of the electrical disturbance accompanying the nerve impulse under these conditions.

The electrical variation passes along the nerve in the form of a wave, so that a simultaneous disturbance may be detected extending over a piece of nerve some 18 mm. long, the crest of the wave being somewhat in advance. The rate at which this wave travels has been often studied, but without concordant results. The determination of this rate in a frog's nerve was one of the early triumphs of von Helmholtz, who found it to be about 30 meters per second. In warm-blooded animals the best observations indicate a rate of about 40 meters per second. Reducing these to expressions with which we are more familiar, these rates are represented by that of a railroad train running at 68–90 miles an hour, or a horse going at the rate of one mile in 40–54 seconds. These figures should be applied with caution, since they are the averages of widely divergent single

[1] Gotch and Horsley, *Proc. Roy. Soc.*, London, 1888.

observations, the cause of this divergence being still undetermined. It has been suggested that the rate was not the same in the different portions of the nervous system, but this suggestion is without evidence, though there is reason to think that in passing through the cell-body, or in passing from one cell element to another, the impulse is much delayed. The signs of metabolism following activity in the nerve fibres are so slight that they have not yet been clearly demonstrated, therefore it is possible to assume that either there are metabolic changes which have not yet been detected, or that the nerve impulse is not accompanied by such changes. However this question shall be decided for the fibres, there is the plainest evidence for metabolic changes in the bodies of the nerve cells, but the description of these will be reserved for the chapter on fatigue. If the nerve impulses are normally initiated at the tips of the cell branches, metabolic changes which have not yet been discovered are to be expected at these points.

To arouse a nerve impulse several conditions must be fulfilled. First a sudden stimulus must be applied. For example, the rapidly interrupted electrical current stimulates, whereas a slow variation in the intensity of this same current does not. Or, again, when mechanical stimulation is employed a slow increase of pressure fails to elicit a response, whereas a light tapping will do so. Gradual changes, therefore, do not act as stimuli. This, however, is by no means equivalent to the statement that they are without effect, although concerning the effect of such inefficient stimuli nothing is known. It has been also shown that the determination of a reaction is more dependent on the rate at which the stimuli occur than upon their strength, and that a single stimulus does not elicit a response. It appears, in consequence, that the arrangements in the central nervous system are such that it reacts more perfectly to a series of slight stimuli than to a small number of strong ones. When in an afferent nerve an impulse is aroused under these conditions it passes on to a cell-body, and there initiates changes which, though in the main similar in different cells, have probably local peculiarities according to the elements involved. These changes cause a second set of impulses, which may leave the

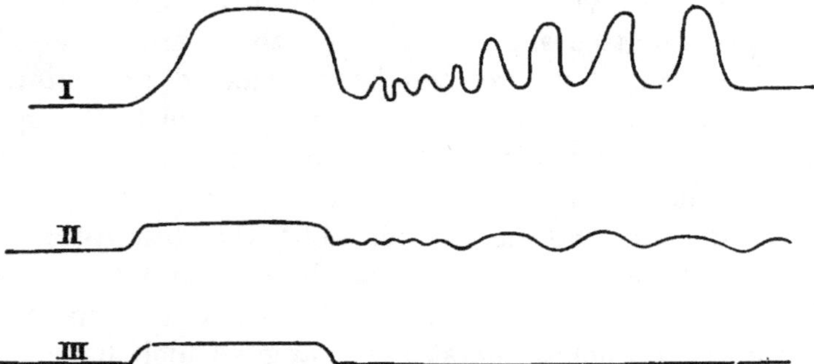

FIG. 63.—A record of the variation in the contraction of a muscle, and the electric condition of the pyramidal fibres upon stimulation of the cerebral cortex by electricity. (Gotch and Horsley.) I, contractions of the muscles, first tonic then clonic ; II, variations in the electrical condition of the pyramidal fibres, as recorded by a capillary electrometer ; III, record of the electric stimulus applied to the cortex. The duration of the stimulus is measured by the length of the elevation in the record.

cell-body with a rate and force quite different from the first. The rapidity with which the impulses follow one another has been recently studied with success. When during an epileptic attack of cortical origin the muscles of a limb are thrown into contraction, they are first held in a tonic spasm, which in turn gives way to a clonic series. The record of the muscular contractions during such an attack would be similar to that expressed by the accompanying curve (I), Fig. 63. The curve (II), however, is the record of the manner in which the nerve impulses followed one another along the spinal cord as the result of stimulating the cerebral cortex, and curve III the record of the stimulus. Since the results occur when the cortical cells alone are stimulated, it is inferred that the peculiar character of the contractions during the epileptic attack depends upon the manner in which the cortical cells discharge. Looking more closely at this reaction, it appears that the discharge begins almost as soon as the stimulus is applied to the cortex, but it may nevertheless go on for some time after it has been withdrawn. The discharge, therefore, takes more time than that occupied by the application of the stimulus; moreover, it is not continuous, but rhythmical.

On anatomical grounds the nerve cells are considered separate units, between which not only is there no genetic continuity, but as a rule not even a secondary connection as close as that between the nerve and muscle elements. For this reason it is difficult, strictly speaking, to trace the pathway of a nerve impulse, since now and again there is a gap in its course. As a matter of fact the impulses cross this gap, though the manner of crossing is still unknown. This difficulty has been mentioned in an earlier chapter, and is again stated, since at this point it plays an important part. The cell outgrowths are the channels along which the waves of nerve impulse are guided in their passage through the central system. Repeated incoming impulses produce such changes in the cell-bodies that finally some of them discharge with a force and rhythm of their own, and once taken up by the central cells, even slight stimuli diffuse themselves over the entire central system, although, in any given case, the reactions which we anticipate and record may come from only a limited portion of it. In thus picturing the entire nervous system as a sensitive mechanism, it is evident that it must respond to the surrounding stimuli as does the water of a lake to the breeze; and such is the relation between the central system and its environment that the breeze is always blowing and the waves of change always chasing one another among the responsive elements. If there are no waves then the cells are dead. The breeze still blows, but it falls on a frozen surface, on cells chilled and rigid beyond the power of response.

The ceaselessness of this stimulation cannot be presented too strongly, because those stimuli which do not come clearly into consciousness are but too readily neglected. Yet the responses of this ever sensitive system reacting to never-ending stimuli are by no means always part of our conscious life, and hence these changes must be indirectly studied if they are to be recognised at all. That such variations may be due either to changes in the exciting stimuli, or to different degrees of responsiveness on the part of the central system, is self-evident. The stimuli during the day are many and strong, but few and weak at night, different according to the seasons of the year, and dependent on changes both without and within the body, changes involving not only alterations in those forms of energy for which special sense organs exist, and which produce the sensations of light, sound, taste, odour, and touch, but also in those for which there are no such organs, as humidity, electric tension, atmospheric pressure, and the like. Yet to all these stimuli responses are made, and they are never twice the same.

The changes in the organism as a whole, for the most part unrecognised, are well illustrated by the variations in the capacity for the performance of voluntary muscular work. It has been found by Lombard [1] that this

[1] Lombard, *Journ. of Physiol.*, 1892.

power was decreased by a decrease in the atmospheric pressure, while in summer, several days of high temperature, especially with great humidity, were followed by a distinct loss. The variation in the capacity for muscular work is mainly due to changes in the nervous system, and hence may be taken as an example of the susceptibility of this system to influences for which we possess no special organ of sense. Such general stimuli as have just been mentioned, humidity and pressure, affect all parts of the central system at the same time, yet the power of causing a general effect is common to all stimuli, be they never so localised at their origin.

On the wide diffusion of impulses depends the fact that a slight stimulation of one part of the system is favourable to the reaction following the stimulation of another. Indeed a condition of diffused stimulation is an essential of the waking state, and if it is not forgotten that stimuli which when of low intensity can thus reinforce one another, may become antagonistic when their intensity is raised, no confusion need be introduced by the observations to be cited later. This reinforcing influence of two stimuli can be illustrated by the studies which have been made on the knee-kick.

If the thigh be supported, so that the leg can swing freely, and the tendon below the knee-pan be sharply struck, the leg will be kicked out. This reaction is brought about by the contraction of extensor muscles in the thigh, and requires for its normal occurrence both the nervous connections of these muscles with the cord, and also a healthy condition of the cord. The study of the normal knee-kick has shown that it differs widely in different persons ; that in the same person it varies from day to day ; that during sleep it may disappear, while in the waking state it may show not only rhythmic variations, but a vast number of irregularities, some of which can be explained.

For example, let the subject gradually fall asleep, the knee-kicks become less and less extensive, as can be seen by the accompanying graphic record; then suddenly it becomes vigorous again.[1]

This revival may be caused by any sort of stimulus—the passing of a heavy cart, the dropping of a pencil, a

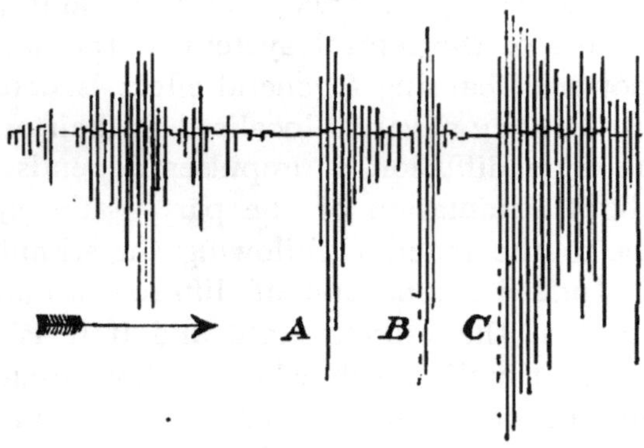

FIG. 64.—Record of the knee-kick of a demented patient. The knee was tapped at regular intervals of five seconds. While the patient was asleep and all about was quiet no response was obtained. After such an irresponsive period the sound of some one walking on the floor below caused at *A* a series of kicks, which gradually diminished, the same at *B*. At *C*, two taps with a pencil and a distant locomotive whistle produced a longer series. The arrow indicates the direction in which the record is to be read. (Noyes.)

flash of light, or irritation of the skin. In each case the result is an increase in the extent of the knee-kick, as shown by the increased length of the verticals in the graphic record[2] (Fig. 65).

The feature in these results most important for us is this—the nerve cells controlling the knee-kick are located

[1] Noyes, *Am. Journ. of Psychology*, 1892.
[2] Lombard, *Am. Journ. of Psychology*, 1887.

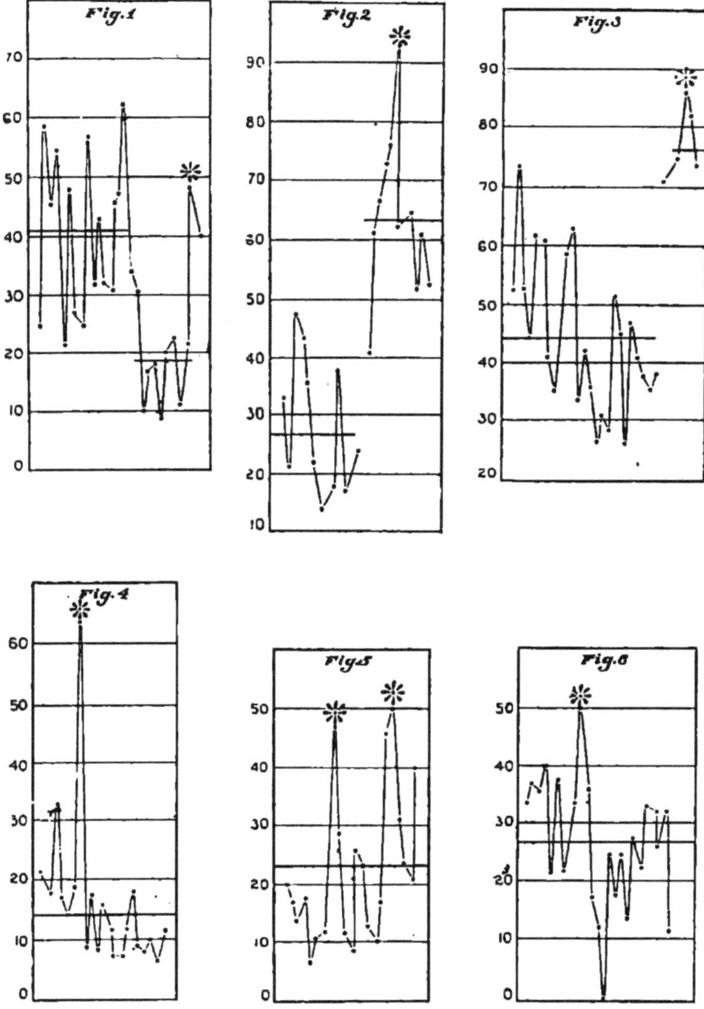

FIG. 65.—A series of small figures showing various reinforcements of the knee-kick. (Lombard.) Fig. 1, subject asleep, when the curve is lowest; * kick after being called. Fig. 2, first part of curve low; * increase in the knee-kick on repeating Browning's poem, "How they brought the good news from Ghent to Aix." Fig. 3, * reinforcement as the result of talking. Fig. 4, * reinforcement due to itching of the car. Fig. 5, * reinforcement due to the crying of a child in the next room. Fig. 6, * reinforcement due to swallowing.

in the lumbar region of the cord. Yet the various stimuli which have been enumerated have their points of entrance into the central system not only at different localities, but at those far removed from the lumbar region; nevertheless in every case reactions due to impulses leaving the cord in this region are affected. The diffusion of the incoming impulse is beautifully shown by the studies on the quantity of blood in a limb during sleep (Bardeen and Nichols).

These experiments were made as follows:—The arm was encased in a plethysmograph, a glass cylinder filled with water, and so arranged that the quantity of water within it varied as the limb swelled or shrank. These variations in the volume of the arm were recorded as given in Figs. 66, 67. During sleep the arm normally swelled. The curves begin at the left, the arm being in a swollen condition. When the sleeper was disturbed by a sound, though he was not awakened, yet the nervous system was aroused sufficiently to give a vigorous response, and the volume of the arm diminished by the withdrawal of blood. This is indicated by the depression of the curve. The value of these experiments for the present discussion lies in the fact that they are typical, and there is warrant for assuming that, if other forms of stimuli were taken and their effects measured by still other reactions, there would be found the same indication of diffusion. The consideration of these facts is important, especially by reason of the apparent opposition in which they stand to the doctrine of localisation, with which they cannot fail to be contrasted. By way of harmonising them, it should be remembered that the facts of localisation are best demonstrated by the reactions of efferent cells, whereas the phenomena of diffusion pertain to the afferent and central elements. They are two opposite aspects detectable in all complete reactions, and must be regarded rather as mutually influencing than as excluding one another. Such being the case in respect to the diffusion of impulses, it becomes important to examine the physiological status thus created by all those impulses entering the central system

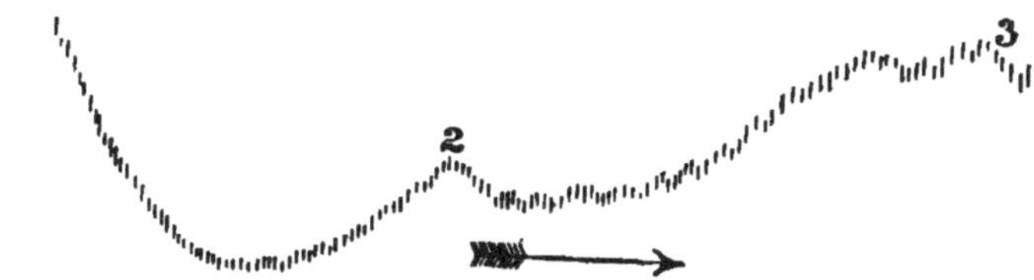

FIG. 66.—Plethysmographic record taken from the arm of a person sleeping in the laboratory. A fall in the curve indicates a decrease in the volume of the arm. The curve is to be read in the direction of the arrow. 1, the night watchman entering the laboratory; 2, the watchman spoke; 3, watchman went out. These changes occurred without awakening the subject. Permission to publish these records was granted by Prof. W. H. Howell. The experiments were made by Messrs. Bardeen and Nichols, students at John Hopkins Medical School.

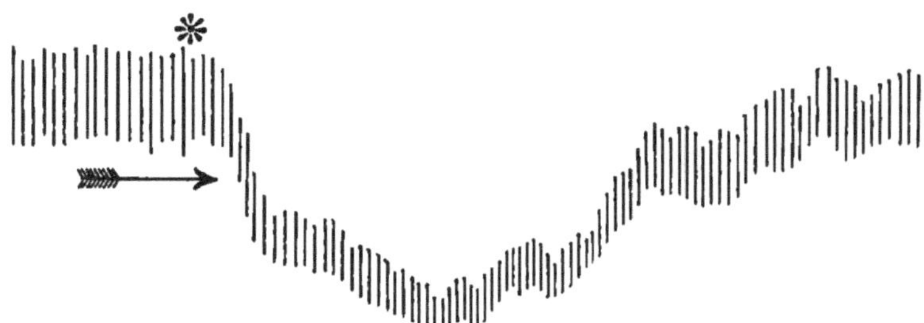

FIG. 67.—Record similar to that in Fig. 66. Change in the volume of the arm of the sleeping subject caused by the sound of a music box which was started at *.

at any moment, since this status forms the background upon which reactions following fresh stimuli must be developed.

The evident stimuli can be measured or estimated,

but the condition of the nervous system on which these stimuli play is ever changing and never capable of any but the most general expression; hence the difficulty of predicting the reactions of the organism as a whole. It is possible, however, to study some of the general changes which occur in this background, and it was as a foundation for this study that the observations bearing on the diffusion of the nerve impulse have been introduced.

Granting that the central system responds throughout a large extent to all the impulses acting upon it, and that by virtue of these responses the background varies, it remains to be determined how far the law and order in these changes can be formulated. There are some observations on the reinforcement of the knee-kick [1] which are important in this place. In these experiments it was found that, if a voluntary contraction of the hand was made by the subject about the time that his tendon was tapped, the extent of the knee-kick was thereby influenced. The modification thus introduced by the additional activity had the effect of increasing or decreasing the height of the kick according to the time relations of the two events. The following diagram expresses the results of varying this relation (Fig. 68).

This curve is typical, and although there is some variation in the reaction according to the nature of the stimulus and the peculiarities of the subject, yet in general the cord reacts more vigorously just after the reinforcement.

The reactions obtainable by the application of a given stimulus depend on the other stimuli with which the new one is competing. Cold water does not feel cold after ice; a black line on a grey surface has a value

[1] Bowditch and Warren, *Journ. of Physiology*, 1890.

different from the same line on a white one, and so on throughout all contrasts.

This relation between the stimulus and the sensation is expressed in the psycho-physic law by the formula that sensation increases in intensity according to the logarithm of the stimulus. The limits within which this law is applicable do not immediately concern us, its use here being merely to express the fact that at any moment

FIG. 68.—Showing in millimeters the amount by which the "reinforced" knee-kick varied from the normal, the level of which is represented by the horizontal line at o. The time intervals which elapsed between the clenching of the hand, which constituted the reinforcement, and the tap on the tendon, are marked below. The reinforcement is greatest when the two events are nearly simultaneous. At an interval of 0.4" it amounts to nothing, during the next 0.6" the height of the kick is actually diminished the longer the interval, after which the negative reinforcement tends to disappear, and when 1.7" is allowed to elapse, the height of the kick ceases to be affected by the clenching of the hand. (Bowditch and Warren.)

the activities of the nervous system under the influence of existing stimuli form a background against which a new stimulus according to its intensity and character may or may not be recognised. Pierce and Jastrow[1] have reported observations bearing on just this matter. Briefly these interesting experiments were as follows:— Two illuminated surfaces were compared when the intensity of the illumination differed by a very slight yet measurable amount, the subject being required to state which surface was the brighter. The difference was so slight that it could not be recognised, and the subject was therefore compelled to "guess." The result of "guessing" showed that the brighter was correctly designated with a frequency so great that the unrecognised difference was clearly effective in determining the choice. The observations have shown that differences too small to be discriminated may still influence our reactions, and it is thus seen that among effective stimuli there must also be included those which we do not recognise.

[1] Pierce and Jastrow, *Mem. Nat. Acad. Sci.*, Washington, 1884.

CHAPTER XV.

PHYSIOLOGICAL RHYTHMS.

Dependence of the central system—Distribution of blood—Physiological rhythms—The greater cycles—Rhythms of the species and the individual—Long rhythms—Daily rhythms—Artificial rhythms—Variations in muscular power—Minor rhythms—Pathological variations in reaction—Automaticity.

UP to this point the nervous system, and the changes which occur in it, have been treated as though isolated and independent, and no effort has been made to emphasise the position of this system as one of several mutually dependent upon each other for their common existence. Nevertheless, the brain and cord are at the mercy of the circulating blood, and through the distribution and quality of the latter the vigour and speed of their reactions are entirely controlled. The central system thus stands in the relation of an engine to its fuel and furnace. The engine may be never so fine, but if the motive power is not sufficient it will work poorly or not at all. So in the animal body, let there be an imperfection in the working of the heart, rigidity or flabbiness in the vessels, a delinquent gland disturbing the nutritive conditions under which the nerves act, and at once the entire system is affected. These facts have a very direct bearing. When, for example, it is a question of inferring the mental abilities of an individual from the gross anatomy of his brain, it must always be remembered that nutrition has been left out of account. In any given case, the abundance of the blood supply to the encephalon can be expressed only in a vague way, as an inference from the size of the heart and blood vessels. Taking such facts into consideration, it becomes intelligible that there should be within us a vast number of nutritive variations, and that between the states of "well" and "ill," which are clearly appreciated, there is a long series of undefined conditions. Many of these changes, dependent both on the influences from without and within the organism, are subject to rhythmic variations.

Of the external stimuli, those of light, sound, and temperature undergo a rhythmic change in intensity corresponding to the alternation of day and night, and such changes are still more clearly marked in the nutritive variations to which the central system is subjected. These physiological rhythms are habits of organic activity which have been found advantageous, and which in man are highly developed. As habits, they are open to modification, but owing to their fundamental character they probably change only with great slowness.

All the lesser cycles of the individual are encompassed by the one great period of growth, beginning with the fertilisation of the ovum and ending with somatic dissolution. Inevitable and rooted in the constitution of the organism, this places the termini between which the other changes fall. Yet there is, perhaps, a cycle beyond this if, taking man collectively, we attempt a wider view, and put the species in the place of the individual. Thus it is not to be forgotten that to anticipate a termination of our species on the earth is but looking forward, as we look backward to its earliest appearance, and within the limits of this greatest period must lie the record of humanity.

History shows that races rise, flourish, and disappear, as do the special civilisations that may spring up among them, and it is probable that all the phenomena of development are as surely to be detected in the species as in the individual and his constituent cells.

Of late, it has been recognised that the living substance composing the higher organisms may be divided into two portions, the body, or soma, and the germ-plasm, or reproductive cells. In the developing animal certain cells are early set aside to form the reproductive elements, and about these the remaining cells build up a protecting body. The great mass of elements which form the soma is, therefore, contrasted with the much smaller mass that forms the germ-plasm, and these subdivisions of the individual have quite different histories. The germ-plasm giving rise through successive generations to new individuals wears the appearance of immortality, while the soma, which shrinks and breaks down after its traditional cycle of "threescore years and ten," has no such claim to distinction. Dependent on slow changes wrought in the germ-plasm must be the long growth period of races and nations to which allusion has just been made, but the rhythms here to be discussed are those occurring within the time limits of somatic life. An analysis of the growth-period shows that in the process of increase in size, phases of rest alternate with those of activity. This rhythm is most rapid at the beginning of development, and probably the vigorous growth at puberty is but the last well-marked acceleration, after the completion of which increase becomes slow, then ceases, and finally gives way to the changes characteristic of old age. Within the life-cycle comes that of reproductive capability, with its beginning at adolescence, its maximum at the completion of somatic growth, followed by a slow decline, leading to cessation in old age.

In the temperate latitudes man is subject to the cycle of the year. This is shown by growth and change in body-weight in the young, and in the mature by alterations in nutrition. But it is in the æstivating and hybernating animals that there is probably to be seen the full development of those changes dependent on the returning seasons—changes which are only suggested in ourselves. Down [1] has observed that in England idiots lose in winter the few acquisitions which they have been able to make during the summer, when there is least tax on their poorly nourished bodies. These seasonal variations, with their long rhythms in nutrition and heat regulation of the body, are again broken into monthly periods. Most marked of these is the menstrual period in the female, with its numerous concomitant changes.[2] In the males, also, there is perhaps a trace of this lunar cycle,[3] while under pathological conditions appear a whole series of periodic phenomena, insanities, and the like, impulses to "spree" and "bad days," which also seem to have a common basis in such a monthly rhythm.

The weekly period is only slightly marked and not organically established. When, however, the daily cycle is examined, many important rhythms become evident, some of which are organically based, but many of them quite dependent on varying social customs.

With the alternations of daylight and darkness there has developed the corresponding alternation of activity and repose. In normal individuals under the most natural conditions this amounts to an organic habit, and where such a habit has been well established, it is only to be slowly broken. On the other hand, variations in external

[1] Down, *Mental Affections of Childhood and Youth*, 1887.
[2] Ellis, *Man and Woman*, London, 1894.
[3] Nelson, *Am. Journ. of Psychology*, 1888.

conditions play a large *rôle*, and the long summer days as well as the long winter nights of the high north tend to prolonged activity followed by prolonged rest. Some happily constructed individuals can sleep for short periods, and thus refreshed take up their work, being fortunately independent of the place or hour. We recognise this capacity as especially developed in the animals whose principal sense is that of smell or hearing, and to whom sleep is an ever-present resource, quite independent of the time of day; while the birds dominated by vision are more strictly controlled by light, and, with the exception of the nocturnal or crepuscular species, in which the conditions are reversed, have their repose determined by the darkness. Yet even in birds during periods of excitement, as when breeding or migrating, the influence of surrounding conditions may be defied. On the other hand, birds mistake the darkness of an eclipse for the evening, and return to roost. Thus, among different groups of animals are to be found the same wide variations in the rhythm of rest and activity as among men; at the same time the fact is not without interest that the animals depending mainly on the intermittent olfactory stimuli exhibit less marked daily rhythm than do those controlled by the visual stimuli that so perfectly dominate the birds.

Like all rhythms, those of the nervous system demand for their production variations in the stimuli or in the condition of the system itself, or both. The least serious changes are those caused by modifying the incoming stimuli, whereas modifications of the physiological condition of the nerve centres are more fundamental. Normally, changes of the latter sort are brought about by variations of the blood supply due to organic reflexes, but to effect them drugs also are often used. It makes no difference whether excitants or depressants are employed, the difficulty for the organism is the same, in that changes are wrought in the central system which cannot be maintained without the repetition of the drug. Stimulation of the cells forming the body is a normal and necessary process, and what it has been agreed to call food acts as a stimulant, but with the minimum damage to the elements. By contrast the group of substances which are classed as drugs produce more decided effects and apparently do more harm. The real difficulty in properly utilising the rhythms produced by drugs depends on the fact that while the normal rhythms involve a mutually dependent variation in several systems, the drugs tend to

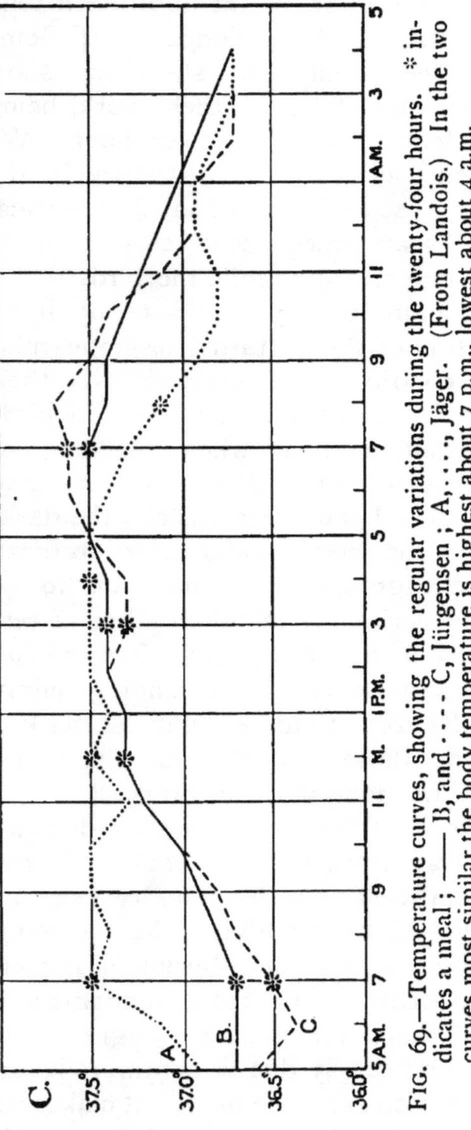

FIG. 69.—Temperature curves, showing the regular variations during the twenty-four hours. * indicates a meal; —— B, and - - - - C, Jürgensen; A, ... , Jäger. (From Landois.) In the two curves most similar the body temperature is highest about 7 p.m., lowest about 4 a.m.

have an unequal action, and are highly selective as to the system which they affect. Thus a physiological reaction which, to be normal, demands a subdivision of labour, is assigned in too large measure to one system, which, being unsupported by its neighbours, ultimately suffers harm.

To return now to the question of daily rhythms in the healthy adult. From morning to evening stature varies; in the morning the stature may be 1 cm. greater than at night. Perhaps the inability to hold one's self erect has something to do with this, or it may be the result of the general shrinkage of the intervertebral cartilages, for it is in the sitting height that the chief loss occurs. During the day the body-weight increases, the curve being the resultant of several factors, the ingestion of food and the processes of excretion. Along with this goes a regular rhythm in the bodily temperature, as shown in the curve below.

This curve is fairly well established, but like so many of the other rhythms, it is partly the outcome of the habit of work, and three meals during the day, followed by rest and fasting at night for observation, indicates that those who work at night and rest during the day have this rhythm reversed.

Several observers have noted a curious periodicity in the capability for muscular work. In one series of experiments the results were the following:—On repeatedly raising a weight by the flexion of a finger, it was found that as the finger tired the weight was raised through a smaller and smaller distance, until finally the muscles altogether failed to respond to the strongest voluntary impulse that could be brought to bear on them. If, despite this failure, the effort to raise the weight was regularly repeated, the contractions reappeared and finally resumed their wonted power, so that the work done by a single contraction was almost as great as before the loss of power. The following curve represents several such periods, with the loss and return of power.[1]

Since the periods do not occur upon direct stimulation of the muscle, the loss of power is probably due to changes in the nervous rather than the muscular system, and the loss was therefore taken as the first indication of central fatigue. By an automatic arrangement the work which was done in these experiments

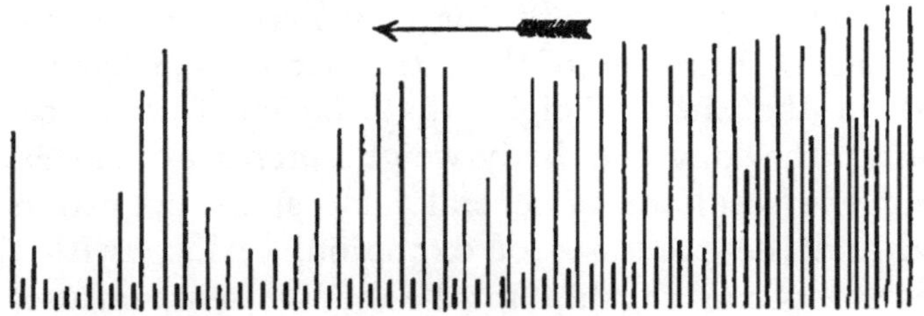

FIG. 70.—A record of the extent of the flexions of the forefinger. The light lines are those for the voluntary contractions. The heavy lines, those for contractions following the direct stimulation of the flexor muscles by electricity. In the former there are periods, in the latter none. The arrow shows the direction in which the curve is to be read. (Lombard.)

was readily determined. When, therefore, it is stated that there is a regular variation in the capability for voluntary muscular work in the course of twenty-four hours, the measure used in making this determination is the amount of work accomplished between the beginning of a record and the first evidence of central fatigue, as shown by the failure to obtain a voluntary contraction. In this way it was that Lombard determined the onset of fatigue at different hours of the day.

It appears that this capability for work was lessened by general fatigue, hunger, declining atmospheric pres-

[1] Lombard, *Journ. of Physiol.*, 1892

sure, high temperature, and tobacco; it was increased by practice, rest—especially sleep—food, increasing atmospheric pressure, and alcohol. In this we have an instance of the variations in the effectiveness with which voluntary impulses act on the centres in the spinal cord.

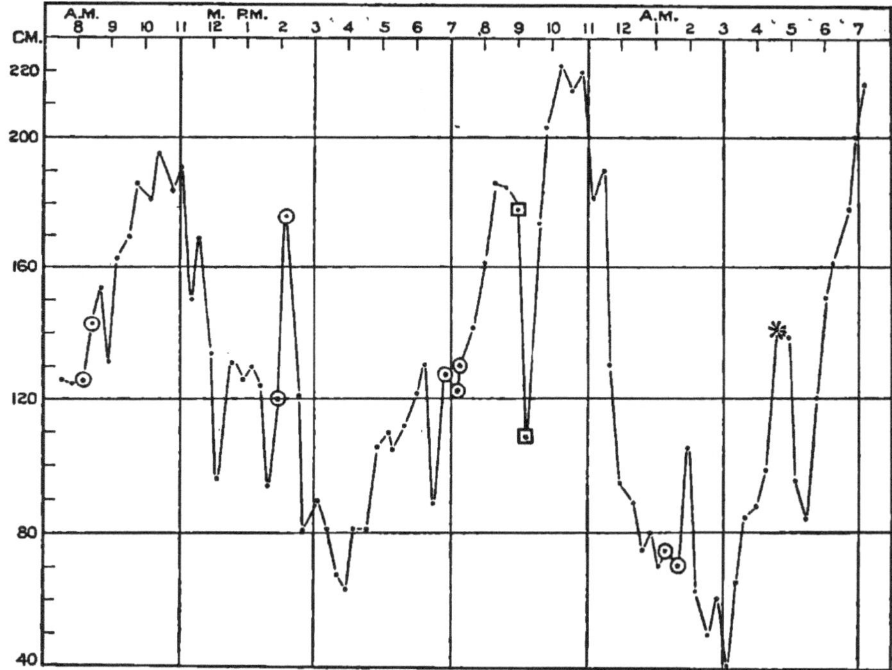

FIG. 71.—Showing at each hour of the day and night how many centimeters a weight of 3,000 grammes could be raised by repeated voluntary contractions of the forefinger before fatigue set in. The curve is highest at 10 to 11 a.m., and 10 to 11 p.m. Lowest, 3 to 4 p.m., and 3 to 4 a.m. Circle with dot, observation made just after taking food; square with dot, smoking; * work done eight minutes after drinking fifteen cubic centimeters of whiskey. (Lombard.)

Recently Ostanikow and Gran[1] have published some interesting figures concerning the diurnal variations of reaction times. When a stimulus is given, and in

[1] Ostanikow and Gran, *Neurolog. Centralbl.*, 1893.

response to it a movement is made, the time elapsing between the first and second event is the reaction time. The figures in Table 57 represent such times, and their diurnal variation is the feature to be illustrated. Of course, between the initial and final events various complicated processes may be intercalated. For example, the stimulus may be used as the signal for a problem in mental arithmetic, or the association of simple ideas, at the conclusion of which the final reaction is to be made, and thus the total time of the reaction is modified by the speed with which this intermediate process is carried on.

TABLE 57.—SHOWING THE LENGTH OF TIME REQUIRED FOR REACTIONS UNDER TWO SETS OF CONDITIONS, AND AT DIFFERENT HOURS OF THE DAY.

The experiments were made at the University of Kasan, Russia, by Ostanikow and Gran, medical students. Tichon was a laboratory servant, eighteen years of age; Galkin, a peasant, sixty-five years of age.

		TIME REQUIRED IN σ. STRAINED ATTENTION.			
Subject.	Form of reaction.	9 a.m.	1 p.m.	4 p.m.	7 p.m.
Gran ...	Simple	167	172	174	171
" ...	Complex	1,116	1,154	1,386	961
Ostanikow	Simple	142	148	181	148
"	Complex	480	460	542	310
Tichon ...	Simple	193	—	214	196
" ...	Complex	1,347	—	1,456	1,217
Galkin ...	Simple	260	242	—	236
" ...	Complex	2,568	2,562	—	2,181

σ = ·001 of a second.

From this it appears that under the conditions of the experiment, the reaction time, both simple and complex, tended to be longest in the afternoon and shortest in either the morning or the early evening. The simple reaction times are shortest in the morning, whereas the others are shortest at the end of the day.

The determination of a diurnal rhythm for purely intellectual work is complicated by the progressive fatigue of the nervous system as the day advances, and to a large extent by external conditions. It certainly depends on food supply, as all experiments indicate, yet digestion and brain work which call for the maximum blood supply at two different localities are not to be carried on simultaneously. It is difficult to get statistics bearing on the processes in question or to measure them ; for everything, from the orderly marshalling of the mental images in a logical line, to the patient waiting for isolated ideas to take shape, can be classed as intellectual work, and in the face of any statement which might now be made, there arise the shades of dead authors to affirm that it was far otherwise with them. For the present at least the problem must be left as one too complicated for general expression.

Of the shorter rhythms there are many. As a rule there are three periods of digestion and absorption corresponding with the three meals, associated with changes in the circulation, heat production, muscular strength, and the emotional and mental activity, all of which are evident, even in spite of physiological arrangements, the assumed purpose of which is to maintain uniformity in these processes. There are, too, the long slow rhythms of blood pressure extending over many seconds, accompanied, perhaps, with corresponding variations in the excitability of the central nervous system, while as smaller waves on this ground swell appear the oscillations due to the respiration and the heart beat, all of which are in turn interdependent.

Passing to the reactions of the nervous system, there

may be added those rhythmic variations which occur when we attempt to fix our attention steadily on one point, as well as those in the muscles of expression, variations which are recorded in everything from the rhetorical period to tremor. These variations in the bodily condition are in general more marked in women than in men,[1] and the manner in which our entire existence is shot through with them is, in the light of the observations given above, amply evident.

In view of this, the times and seasons for the tasks when the exercise of the nervous system is undertaken are worthy of careful study, for on the proper selection of them will depend in some measure the results.

In popular estimation the organism is looked on as subject to but little change, while the surrounding conditions are made responsible for the differences in reactions that are noted in different persons or in the same person at different times, yet as a matter of fact the variations in the organism play the more important *rôle*. Though in health we are for the most part delightfully unconscious of these changes in ourselves, in abnormal conditions the consciousness of them can be greatly modified, and that too in either direction. A patient may be in bed for months with all the sensations of a broken leg, when there has been no break; suffer from constipation, through neglect of the ordinary visceral sensations, which are somehow disregarded; or have for all practical purposes a paralysed arm which has grown out of the fear of pain, should the attempt be made to move it. Instances of this sort lie so close to every-day experience that a case may be quoted:[2]—"A lady about forty years of age, the wife of a physician, consulted me in September, 1885, about her left shoulder;

[1] *Vide* Ellis, *Man and Woman*, 1894.
[2] Taylor, *Journ. of Nerv. and Ment. Dis.*, 1888.

she had wrenched it three months before while trying to save herself from falling on the stairs. She did fall and bruised herself in several places, but not on the shoulder. Her arm was afterwards stiff and painful, and she found it powerless at the shoulder and elbow; she carried her arm in a sling, and it had been treated electrically. At the time of the examination there was pain in the elbow when she raised the arm more than 45 degrees from the side. When passive movements were made the muscles about the shoulder resisted, and motion was not free. Diagnosis of restraint and disturbed reflexes. Training of the reflexes by systematic movements was followed by marked improvement in mobility and usefulness of the arm, but after being treated for a week the patient was obliged to leave, and went home with the arm very much disabled. It remained in about the same condition until the death of her husband, which occurred unexpectedly a few weeks later. The shock was so great that she became entirely unconscious of her arm, and from that it has been perfectly normal in every respect, as she was able to prove to me at her next visit." In this instance by a twist of attention the patient became unduly conscious of processes in her spinal cord, which were, so to speak, none of her business, and it required another twist to free the spinal centres from the abnormal surveillance, and restore the functions of the muscles.

Before leaving the subject in hand the perplexing question of automaticity in the nervous system must be touched upon. Examples of automaticity are the movements of an amœba, the pulsations in the umbrella of a jelly-fish, the beat of the heart, the rhythms of respiration or the purposeful movements of a higher animal. In these instances, with the exception of the amœba, we have the reactions under the control of a distinct nervous system, and the question takes the form whether the nervous elements can discharge without being immediately excited by incoming stimuli.

The test would be found in the reactions of these cells when deprived of all such stimuli. As a result of attempts to do this it is discovered that when the stimuli can really be removed the reactions cease, and that the nonconforming cases are those in which the stimuli cannot be controlled. As applied to the movements of the entire animal the view that automatic reactions are but a series of responses to stimuli more or less masked, may be illustrated in the following way: If a frog be deprived of its brain (encephalon), the cord being left intact, it will remain irritable for several days. Such a frog exhibits no automatic movements. It remains where it is placed. If the skin is stimulated, it reacts by a contraction of the appropriate muscles, and the reaction being completed, relapses into the quiescent state. The force and extent of the reaction in such a case is predictable since it depends in a large measure on the character of the stimulus. Let such a frog be compared with others in which successively smaller portions of the brain have been removed. As the quantity removed becomes less there is an increase in the number of the sense organs remaining intact, together with an increase of the total mass of nervous centres. The less the quantity removed the less predictable become the reactions of the frog, and the more difficult is the correlation between them and the stimulus, because for one thing there are greater variations in the excitability of the nerve centres themselves. But at best a normal frog is not a very complicated affair. It can safely be said that when pursued it will try to escape, though just how many jumps it will make and in what direction it will go are uncertain. If, passing up the zoological scale, a bull-dog is irritated, the prediction is more difficult. He may try to escape or he may try something else ; yet whatever he does he will in most cases react shortly after the application of the stimulus. Stimulate a man and there are possible these forms of reaction with endless elaborations and modifications, but there is also a possible exaggeration of the time which may elapse between the application stimulus and the response. The interval may be only so long as is required to deal a blow, or the man may bequeath his response to an insult to his sons and establish a family feud. The capacity for postponing a reaction is the mark of the more highly developed animals, and especially of man. Often when a reaction takes place it seems to have no apparent relation to the stimuli acting at the moment : in one sense it does not have any, while in another it does. For this reason we have among the higher animals that remarkable disconnection between the stimulus of the moment and the response to it, which gives the appearance of a mutual independence, and has thus given support to the conception of automatism in physiology.

In the last analysis, however, the incoming stimuli appear as the causes of all our actions. The chain of events is often long and hard to follow ; but so much evidence stands on the side of the reflex nature of all actions, that the hypothesis of automatism is not longer necessary.

The facts which have been reviewed show that in the nervous system there are changes continually going on, changes which are often rhythmic, and which, because they are easily diffused, affect the system as a whole, just as surely as the displacement of any member must affect the motions of a planetary group. Yet it is only here and there, or now and then, that this ceaseless flux rises into consciousness and reveals to us some notion of our own plasticity.

The rhythms which occur mark periods of greater alternating with those of less activity ; the former being those during which energy-producing substances break down, so that the organism is fatigued, and in this process of building up and breaking down is to be found the basis of those alterations in the nervous background which have been here presented. Fatigue is therefore worthy of careful examination, since a knowledge of the processes causing it can on the one hand suggest why our capabilities periodically diminish, and on the other how this impressive change, the shadow of on-coming death, can be made least harmful.

CHAPTER XVI.

FATIGUE.

Expenditure of energy—Possible increase of potential energy—Results of anatomical disproportion—Fatigue affects the material stored at any one time—Results of exercise—Fatigue a function of the central system—Beneficial exercise—Fatigue substances—Their diffusion—Auto-intoxication—Drugs—Sleep—Relation to central system—Changes in the circulation—Rhythms—Dependence of sleep on stimulation—Variations with age—Depth of sleep—Recuperation—Length of the period—Hodge's experiments—Observations of Mann and Vas—Fatigue of the nerve fibres—Starvation—Other tissues—Résumé.

IT is conceivable that two individuals of equal age and size may have expended during their lives very different amounts of energy. Further, this expenditure need not in every case correspond with effective performance, for internal resistances may absorb a large share of the energy set free. That individuals differ in their everyday performances is recognised, and these inequalities depend both on a more happy organisation of those most capable, as well as on fundamental differences in the capacity for storing energy. Individual differences in these respects include the minutest details, and are remarkably persistent.[1] For example, Mosso found that in the periodic loss of muscular power following prolonged activity, the two individuals whom he specially

[1] Mosso. *La Fatigue Intellectuelle et Physique*, Paris, 1894.

studied showed marked differences. Five years later another set of records showed the same differences. Similar illustrations might be indefinitely multiplied. At any moment the active body-cells contain a mass of substances capable of producing energy by their decomposition, yet the mass present at any moment is but a trifling fraction of that which will be built up and broken down in the course of a normal lifetime. This capacity to build up food-stuffs into its own substance is the peculiarity of the living cell, as contrasted with a crystal or a machine, and it is a question of great interest to learn whether in ourselves, by our own foresight, and within the limits of a lifetime, it can be increased. Thus it would be of importance could we determine whether the greater store of available energy in those who seek by hygienic care to cultivate their effective force is to be counted as an actual addition to their powers, or whether it is simply that fraction of inherited energy which has been rendered available by more effective expenditure. Probably both changes follow. Yet, so seldom are the natural forces normally expended, that it is difficult to get the data for a judgment.

In the body-form deviations from the mean are familiar, and their implication appreciated. The build of the "sprinter" is something quite different from that of the lifter of heavy weights. Or, taking deviations which are more marked, any part—head, arms, legs, chest—may be too large or too small for the other portions of the body, and it is abundantly evident that these failures in proportion, if those of excess, may easily reach the point where the offending part becomes too large to be correctly nourished or controlled by the remainder of the body; or, if those of defect, where the part must work under excessive strain, in order to contribute its share to the total activities of the organism. Yet these anatomical variations have their physiological counterparts in every direction. A man condemned by his profession to sedentary habits may find a splendid muscular system acting upon him as a veritable parasite, or conversely a bulky nervous system, coupled with too small a heart, may lead to an entire mental existence on half rations. In some small measure these maladjustments can be controlled, and by such control the effective activity increased.

In the study of fatigue it is the changes in the material stored in the active cells at any one time that claim attention. Normally exertion precedes fatigue. When the muscles are exercised with the purpose of increasing their strength, or the accuracy of their coordination, their reactions do not improve with regularity, or for an indefinite period. For the first two days or so of systematic exercise the reactions may be fairly satisfactory; then follows a period of poor performance which, after due time, gives way to improvement that comes in waves. This indicates that the first general result of exercising the body-cells, and specially those of the central system, is not an immediate increase in their powers of metabolism, but that this power is only slowly improved.

Even when cut off from their normal surroundings, as when a muscle or part of the spinal cord is isolated, the cells are for a time capable of doing work, but under such circumstances soon become exhausted. If the decrease in power under such circumstances is measured, it is found that it steadily diminishes with each discharge. When a muscle tested in this way has been brought to the point at which it will no longer respond, washing it out by passing a normal salt solution through the vessels will cause a return of power.

Under natural conditions the blood is always streaming through the active tissue, and thus not only removing the waste products, but supplying fresh nutritive materials. In some measure the constructive and destructive processes occur together, but the curve of normal fatigue indicates that the loss of power is first rapid and then slow, exhaustion being long deferred.

The neuro-muscular reactions are under the control of the central system, and it is of interest to recall that, as shown in the last chapter, it is the variations in the nervous system which are most important for them. So, too, in the last stages of extreme fatigue, it is the nerve cells, not the muscles, which are exhausted.

The exercise of cells finally causes both their better irrigation by the blood, and the greater and more rapid storage of fresh material. In this respect the structural element stands related to the nutrient lymph as does the entire organism to its food supply; a good appetite and vigorous digestion indicating health. Closely connected with these processes is the determination of the point to which exercise may be carried, in order that it shall be followed by the maximum increase in power. If the cell activities are viewed as dependent on the partial decomposition of complex molecules, then it is conceived that this decomposition, when carried to a certain point, leaves the cell best prepared to recuperate. This optimum can, however, be overstepped, and instances where persons, after one form and another of excessive fatigue, have never completely recovered, are too familiar to require specification. Just what this limit is, must be determined by personal experiment, always keeping in mind that the individual differences are very wide.

Both the sensation of fatigue, and in part the loss of power, are regarded as due to substances resulting from the decomposition of the cytoplasm during activity. The signs of fatigue can thus be induced in a dog at rest by injecting into his vessels the blood of a dog that has been wearied (Mosso). The substances thus passed into the blood are not only deterrent to the activity of the cells which produce them, but to the other cells with which they are brought in contact by the circulation. Thus local exercise produces general weariness. Moreover, these substances are probably different, and have a different physiological value according to the tissues in which they originate. The fatigue from nervous activity differs from that due to muscular exertion or over-eating. When some error in the elimination or chemical destruction of these substances occurs, there may develop a genuine auto-intoxication the effects of which can apparently be wide-reaching.[1]

Healthy weariness is followed by sleep—a condition hard to define. Unconsciousness may be produced by a blow on the head, by drugs, extreme cold, hypnotic suggestion, compression of the carotid arteries, and so on : from these conditions normal sleep is distinguished by the fact that it is refreshing ; and so far as it departs from this character, it fails of its purpose. The animals in which the habit of rhythmic sleep is best developed are the birds and higher mammals, and among these it is especially those in which the encephalon is most elaborated that there is a sharp contrast between activity and repose. In these forms the central cells are proportionately well developed, and thus there is some reason to associate this condition with variations in the functions of this group. At the same time, the higher animals have it in common with the lower, that

[1] Cowles, *Neurasthenia and its Mental Symptoms*, Shattuck Lecture, Boston, 1891.

some reflexes are still easily aroused even in this state, while others, like the knee-kick, are lost.

The waste products of activity appear as the prime cause of sleep, and on them is probably dependent the distribution of the blood which has been observed during this condition. The effective blood supply of the nerve centres is increased in the first stages of activity, and diminishes in fatigue (Mosso). It is diminished during sleep also. Any stimulus acting on the sensory nerves during sleep tends to cause a withdrawal of blood from the limbs, and its return to the head as shown in the observations of Bardeen and Nichols previously quoted.

Sleep is more easily induced, the fewer the stimuli that act upon us. Not only, then, the condition of the afferent nerves, but also the responsiveness of the central cells toward the impulses that fall upon them, are to be taken into account. In common practice we reduce stimuli to a minimum as a preliminary to sleep. Strümpell[1] has reported an interesting case of a lad of low intelligence, who by disease had been reduced to one ear and one eye as the sole avenues by which he received external stimuli. When the ear was plugged, and the eye bandaged, he fell asleep with the regularity of a machine.

The parts played by the sensory and that by the central cells vary somewhat at different times of life. In childhood the amount of stored material is small, large at maturity, and small again in old age. Hence the cells would, by reason of this fact, have the greatest capability for work in the middle period. Between childhood and old age there is, however, this difference, that while in the former the non-available substances in the cell are developing, not yet having matured, those in

[1] Strümpell, *Deutsch. Arch. f. Klin. Med.*, 1864.

the latter have in some way become incapable of reconstruction. With this change in the nerve elements goes the remarkable power in the blood vessels to vary their diameter, and thus control the supply of blood. The limits of this variation probably increase to maturity, and in old age are greatly diminished, so that the calibre becomes more constant. The degrees to which the blood supply can be controlled, and the

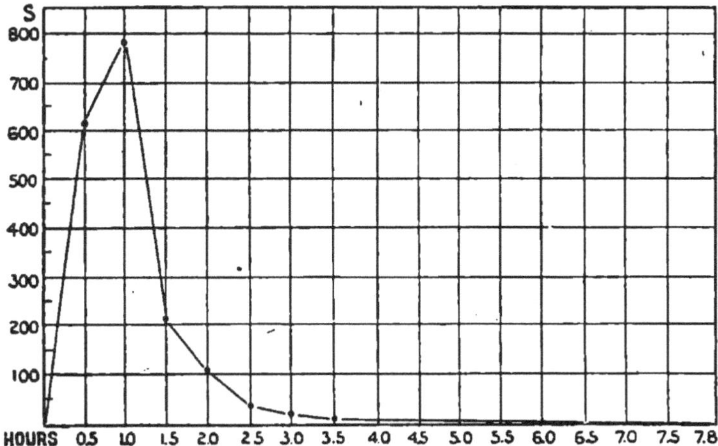

FIG. 72.—Curve illustrating the strength of an auditory stimulus (a ball falling from a height) necessary to waken a sleeping person. The hours are marked below. The tests were made at half-hour intervals. The curve indicates that the distance through which the ball required to be dropped increased during the first hour and then diminished, at first very rapidly, then slowly. (Kohlschütter.)

amount of substance capable of yielding energy at various periods of life are so different, that considering these factors alone, though there are probably others, it may be easily appreciated that the sleep of childhood, maturity, and old age should be quite different.

During sleep stimuli can always produce an effect, otherwise a person could not be wakened. When applied to the cranial nerves the sleeper is apt to be recalled to consciousness by the redistribution of his blood, the supply to the encephalon being increased.[1] Kohlschütter has reported observations on the depth of sleep, which was measured by the height from which a falling ball must drop in order to make a sound sufficiently loud to waken the sleeper.

The ease with which the subject awakens is manifestly not a measure of recuperation, since after two and a half hours of sleep a very slight stimulus is efficient. A condition permitting the diffusion of the auditory stimuli is therefore established long before the reconstructive changes in the cells in general are complete, and there is reason to think that the first change is due to an increased supply of blood to the central system, after which the constructive processes slowly follow in the cells. For this last, a period of four hours or more is usually demanded, and when less time is allowed, the cells return to work under great disadvantages.

The feat of walking one thousand miles in a thousand hours, at the rate of a mile an hour, leaves the pedestrian with at least two-thirds of each hour unoccupied, yet it requires the greatest endurance, from the fact of the discontinuity of sleep; not only, therefore, the amount of time taken for sleep, but the length of the single periods, is significant. Continuous loss of sleep is far more rapidly fatal than starvation, and the final changes are very marked, especially in the nervous system.[2]

These physiological phenomena are correlated with the anatomical variations found in the cell elements at different times. The changes which occur in the bodies of nerve cells as the result of their activity have been

[1] Kohlschütter, *Zeitschr. f. rat. Med.*, 1863.
[2] De Manaceïne, *Arch. Italienne de Biol.*, 1894.

most carefully studied by Hodge, who first demonstrated them.[1] Experiments were made on frogs and

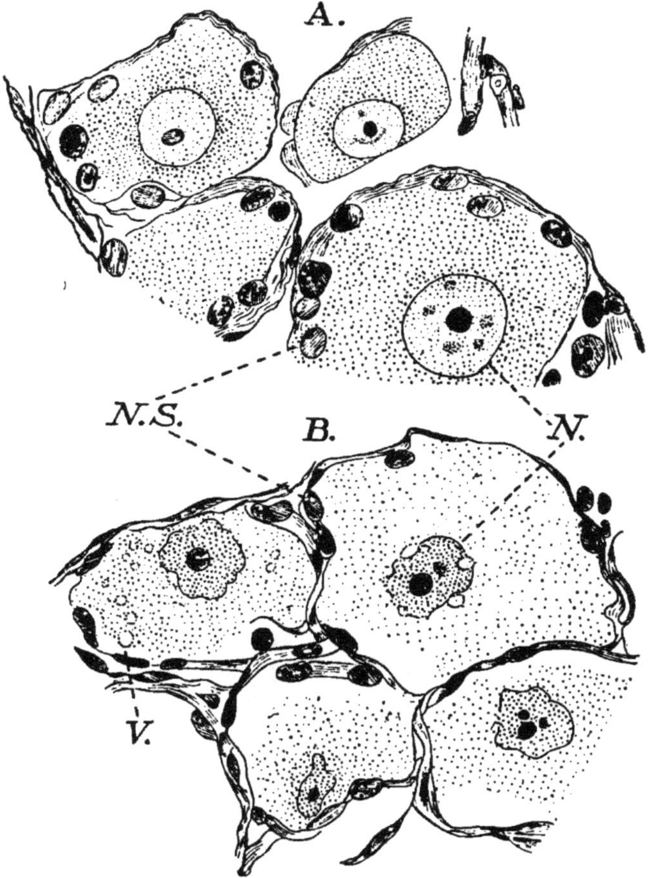

FIG. 73.—Two sections, *A* and *B*, from the first thoracic spinal ganglion of a cat. *B* is from the ganglion which had been electrically stimulated through its nerve for five hours. *A*, from the corresponding resting ganglion. The shrinkage of the structures connected with the stimulated cells is the most marked general change. *N*, nucleus ; *N.S.*, nucleus of the capsule ; *V*, vacuole, × 500 diameters. (Hodge.)

---

[1] Hodge, *Journ. of Morphology*, Boston, 1892.

cats by electrically stimulating the peripheral trunks of mixed spinal nerves; the condition of the spinal ganglion cells, with which the sensory fibres were connected, was then examined; the cells in the ganglion on the side of stimulation being compared with those from the opposite ganglion of the same animal. As a result of stimulation in this way, the fatigued cells appear shrunken, and their reaction to staining reagents changed, thus showing a chemical alteration. The shrinkage involves the nuclei of the capsule surrounding these elements as well as the nucleolus, nucleus, and cytoplasm of the cells themselves, the last becoming vacuolated. In the accompanying figure (73) these differences can be seen by comparing the fatigued cells *B* with the normal cells *A*.

As the most clearly marked differences were found in the nuclei, these were further studied. Their volume decreases according to the length of time that they are stimulated, as indicated in Table 58.

TABLE 58.—SHOWING THE DECREASE IN THE VOLUME OF THE NUCLEUS OF STIMULATED SPINAL GANGLION CELLS OF CATS. FIFTEEN SECONDS' STIMULATION ALTERNATING WITH FORTY-FIVE SECONDS' REST. (*Hodge.*)

STIMULATION CONTINUED FOR	SHRINKAGE IN THE VOLUME OF THE NUCLEI OF THE STIMULATED CELLS.
1 hour	22 per cent.
2·5 hours	21 " "
5 "	24 " "
10 "	44 " "

The curve shown by the broken line in Fig. 74 expresses the same facts as have just been given in Table 58.

The solid line in this figure relates to another set of experiments. Under the conditions there employed, five hours' stimulation was found to reduce the volume of the nucleus almost 50 per cent. A series of experiments was next made in which the stimulation was carried on for five hours, and then different animals allowed to recover for the number of hours indicated, and the condition of the cells examined at the end of the various intervals, extending up to twenty-four hours. The curve shows that there is a gradual return

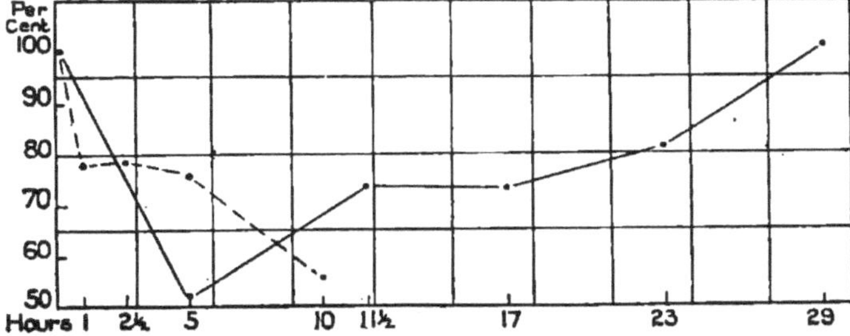

FIG. 74.—The dotted line indicates the volume of the nuclei of the spinal ganglion cells of a cat after stimulation for the times indicated. The solid line indicates the volume of the nuclei, first after severe stimulation for five hours, and then in other cats, similarly stimulated, but subsequently allowed to rest for different periods of time. The actual period of rest is found by subtracting five hours from the time at which the record is made. After twenty-four hours of rest the nucleus is seen to have regained its normal volume. (Hodge.)

of the nucleus to its initial volume. It was thus indicated that the changes in the cell had not overpassed physiological limits.

This result led to the study of those animals in which there were well-marked periods of activity and repose, and in which an analogous difference should be found between the cells of an animal killed in the morning, after a night of rest, and one killed in the evening after a day of exercise. Hodge was thus able to demonstrate fatigued cells in the spinal ganglia of English sparrows; in the large cells from the cerebellum of the swallow; in cells from the cortex of the pigeon and those from the antennary lobes of the honey-bee, the animals being taken at the end of the active day; whereas in animals taken in the morning the cells in these localities had the characters of those at rest.[1]

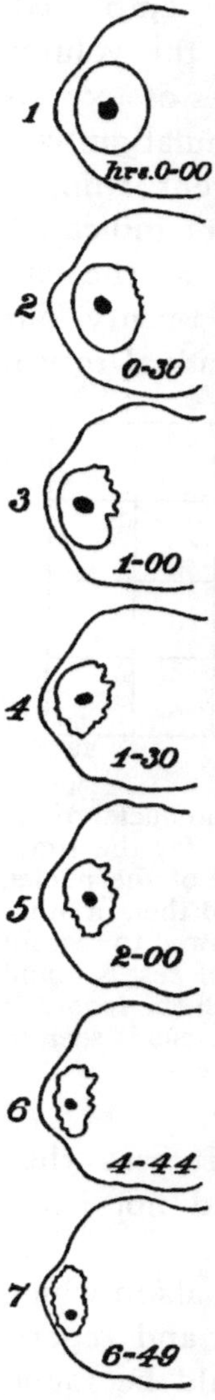

FIG. 75.—Showing changes observed in the nucleus of the living sympathetic nerve cell of the frog, as the result of direct electrical stimulation. The hour of observation is given within each outline. The whole experiment lasted six hours and forty-nine minutes. The control cell treated in the same manner during this time, except that it was not stimulated, showed no change. (Hodge.)

Since these first experiments this same author has been able to put two sympathetic ganglia under the microscope, and to stimulate one while leaving the other at rest, and thus to watch and compare in the living cell the changes occurring during stimulation. In the control specimen no change took place in the cells during the experiment, while in the former, progressive changes occurred, of which Fig. 75 is a representation. This series of experiments presents the first good anatomical alteration to be correlated with the loss of power following exertion; and though thus far they have only casually been applied

[1] Hodge, *Journ. of Morphology*, 1894.

to man, there is little doubt of their complete applicability.

It should be mentioned that some more recent observations by Mann[1] have confirmed the changes found by Hodge after prolonged exertion, and have added the motor cells of the spinal cord and the elements of the retina (dog) to the list of those in which they occur. This author agrees with Vas,[2] however, in finding a preliminary swelling of the cell and its nucleus after a short period of stimulation, the change having been observed in both sympathetic and cerebral retinal cells, and both those authors have added interesting observations concerning the rearrangement of the chromatic substance as the result of stimulation.

Physiologists have been busy at the same time seeking to determine how far the passage of a nerve impulse along a fibre causes fatigue-changes in it. Thermal and chemical changes in the nerve fibres have not been clearly proven. After very prolonged stimulation by electricity, the nerve still remains permeable to impulses thus aroused.[3] Neither have definite morphological changes been satisfactorily demonstrated.[4] This is the more interesting, since the nerve fibre is a direct prolongation of the cell body, and also since other experiments show that the irritability and conductivity of nerve fibres are not necessarily present in the same degree, for a fibre may cease to be irritable, yet still conduct.

In starvation the central system is the one least affected in its gross weight,[5] and there is little doubt that this weight is maintained at the expense of the

[1] Mann, *Journ. of Anat. and Physiol.*, 1894.
[2] Vas, *Arch. f. Mikros. Anat.*, 1892.
[3] Bowditch, *Archiv. f. Anat. and Physiol.*, 1890.
[4] Edes, *Journ. of Physiol.*, 1892.
[5] Voit, *Zeitsch. f. Biol.*, 1894.

other systems—a relation the utility of which is plain, but the mechanism of which is quite unexplained.

In general the fatigue which controls us is, in so large a measure, dependent on the nervous system that there is reason to make the changes occurring there the most prominent, though it can also be shown that the glands and muscles undergo changes as the result of activity, and a complete discussion of the subject would involve all the active tissues.

There is, then, an anatomy as well as a physiology of fatigue, and the facts grouped together permit the following general statement. After recuperative sleep the cells in the central system are full-sized and granular, the blood flows with a medium pressure through the nerve centres; slight stimuli elicit a ready response, and there are general sensations of vigour and well-being. From the beginning of the day the process of running down goes on, all the constant stimuli hasten it, meals retard it, drugs modify it, according to their nature. In general there is a tendency to run down towards the middle of the afternoon, with a return of vigour later in the day. On this long rhythm is superposed one by which in the evening the blood supply to the brain diminishes at the accustomed hour of retiring. This change in the blood supply appears to depend on the waste substances produced by the active cells. These accumulate faster than they are removed, and render activity more difficult. At the beginning of sleep these substances are abundant, the stored material in the cells is small, and the cells themselves are shrunken in various ways. Slowly the toxic products of metabolism are removed, and at the end of two or three hours the sleeper is in a state to be readily awakened, though physiological recuperation has but just begun. The circulation has become better, the constructive changes in the cells continue, and at the end of the interval the nerve cells are restored and the body prepared for the next period of work.

CHAPTER XVII.

OLD AGE.

Changes in the entire body—Change in the weight of the encephalon—In the lobes of the cerebrum—In the thickness of the cortex—In the cerebellum—Studies on paralysis agitans—Observations by Hodge on old age—Influence of specialisation—Effect of exercise—Order of dissolution—Decreasing productivity—Multiple pathways—Similarities between fatigue and old age.

IN advanced life the grosser changes in the body due to growth are amply evident. By the condensation of the skeleton and the greater curvature of the spine the stature decreases towards the evening of life just as towards the evening of each day the body has become shorter. The bodily temperature falls from 0·1–0·5°C. below that found in the prime of life.[1] The weight of the active tissues decreases, that of the cell-multiplying and blood-producing organs being most diminished,[2] but this decrease may be masked by the accumulation of fat, though this also disappears in persons very old. It has been suggested that even the shape of the skull may alter at this time, but sufficient evidence for this view is not forthcoming. In these total changes the central nervous system bears its part. The tables (13 and 29) in Chapters IV. and VI. show that for sane and insane persons of both sexes, the encephalic weight is

[1] Kelynack, *Medical Chronicle*, Manchester, 1892.

[2] Humphrey, *Old Age*, Cambridge, 1890.

least in the group of the aged, 70–90 years. As compared with the 20–40 years' group, the percentage value of the cerebrum is slightly reduced in the sane, the reduction being most marked in the females. Among the insane this variation in the percentage does not occur. The observations by Bischoff[1] indicate that the final decrease in the weight of the encephalon begins in men at about fifty-five years, and in women some years earlier (Fig. 76). Above these curves is plotted that for the eminent men, the list of whose

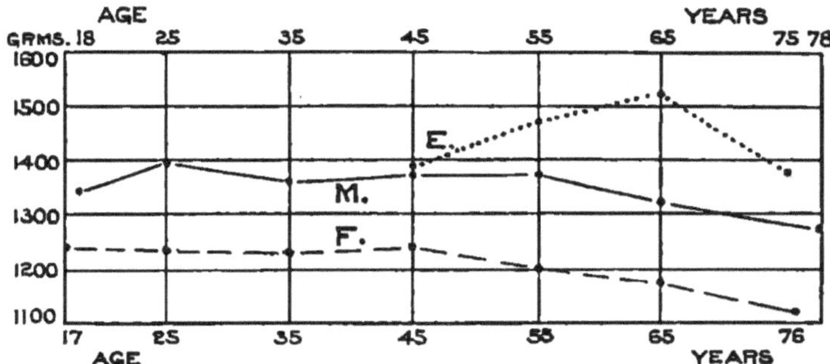

FIG. 76.—Curves for males, ———, females, - - - -, representing the average weight of the encephalon (with pia), are based on the observations of Bischoff. The material was obtained from the less favoured social classes. M, males ; F, females. The curve E . . . . is based on Table 26, and shows that among the eminent men the period of decrease is deferred. (Bischoff.)

brain-weights has been given in Chapter VI., and this curve indicates that among these latter the decline is deferred till after sixty-five years.

This difference is important. The contrast here is between a favoured social class, the eminent men, and the much less favoured class from which Bischoff's data were drawn, and it needs but a moment's thought to see that the favourable conditions might thus tend to defer the onset of decay. This, however, is not the

[1] Bischoff, *Hirngewicht des Menschen*, Bonn, 1880.

whole story. The eminent men have not only the heavier, but the better organised brains; this organisation being dependent in part on prolonged growth, a condition which would also tend to defer senescence. For these reasons the time at which decay commences in the least favoured classes is probably early as compared with the time in those more happily circumstanced. But here again the incompleteness of the data must be noted.

Some effort has been made to determine how the several lobes of the cerebrum are affected by the general diminution in its weight. The tables here given have been taken from Topinard. Each cerebral hemisphere, the pia having been removed, was divided (Broca), as shown in Fig. 30. This gives the subdivisions of the hemispheres, including both basal ganglia and mantle, as presented in the tables which follow.

TABLE 59.—SHOWING THE ABSOLUTE WEIGHT IN GRAMMES OF THE LOBES OF THE CEREBRUM (WITHOUT PIA), ACCORDING TO SEX AND AGE. (*Broca.*)

MALES.	LOBES.		
	FRONTAL.	OCCIPITAL.	TEMPORO-PARIETAL.
25—45 years	502	111	552
70—91 "	429	112	458
Diff. due to age	— 73	+ 1	— 94
FEMALES.			
25—45 years	429	100	482
70—91 "	392	91	416
Diff. due to age	— 27	— 9	— 66

The records show a similar distribution of loss in the two sexes, though it is absolutely greater in the males than in the females. Especially heavy is the loss in the frontal lobes of the males. Expressing these absolute figures as percentages, the weight of the entire encephalon in each sex being represented by 1,000, the results shown in Table 60 are obtained.

TABLE 60.—GIVING THE PERCENTAGE VALUES (NUMBER OF PARTS IN 1,000) OF THE ABSOLUTE FIGURES IN TABLE 59, ARRANGED ACCORDING TO SEX AND AGE. (*Broca.*)

MALES.	LOBES.		
	FRONTAL.	OCCIPITAL.	TEMPORO-PARIETAL.
25—45 years	431	95	473
70—90 ,,	429	112	458
FEMALES.			
25—45 years	424	99	476
70—90 ,,	437	101	462

The proportions of the cerebrum thus divided are nearly the same for both sexes at the two age periods. The most marked difference is in the temporo-parietal lobes, the proportional value of which in both sexes decreases 1·5 per cent. with advanced age. The occipital lobes in the aged are something more than 1 per cent. heavier. In the male, age produces a proportional decrease in the weight of the frontal lobes, whereas in the females there is an increase. Of changes in the weight of the spinal cord nothing is known, neither are there any studies of old age in animals which illustrate the changes in the nervous system.

The thickness of the cerebral cortex diminishes in harmony with the shrinkage of the entire system. In large measure this must depend on the loss of volume in the various fibre systems, which, according to the tables from Vulpius (Chapter XIV.), show a senile decrease in the number of fibres composing them. This decrease is more marked in the motor than in the sensory areas. The time at which it commences cannot, however, be well judged from the curve mentioned, owing to the small number of records after the thirty-third year. Where records are made between this and the seventy-ninth year, it appears that there is no decided diminution until after the fiftieth year, though at the seventy-ninth the decrease is always clearly shown. Engel has shown that the branches of the arbor vitæ of the human cerebellum decrease in size and number in old age. See Fig. 45.[1]

To the anatomy of the human nervous system in old age contributions have been made by studies on the pathological anatomy of paralysis agitans.[2]

In subjects suffering from this affection the bodies of the nerve cells are shrunken, pigmented, and show in some cases a granular degeneration; the fibres, in part, are atrophied and degenerated, the supporting tissues increased, and the walls of the small blood-vessels thickened. These changes have been found principally in the spinal cord, being most marked in the lumbar region. But the cords of the aged persons who do not exhibit the symptoms of paralysis agitans show similar changes, though usually they are not so marked, and hence the pathologic anatomy of this disease resolves itself into a somewhat premature and excessive senility of the central system.

[1] Engel, *Wien. Med. Wochenschrift*, 1863.

[2] Ketcher, *Zeitschrift f. Heilkunde*, 1892; Redlich, *Jahrbuch. f. Psychiatrie*, 1893.

A careful study of senile nerve cells has been made by Hodge.[1] On examining the central system of a man dying naturally at ninety-two years, very marked changes were found in the cells of the spinal ganglia when compared with those of the corresponding ganglion taken from a child at birth. The principal differences are summarised in Table 61.

TABLE 61.—SHOWING THE PRINCIPAL DIFFERENCES OBSERVED ON COMPARING THE SPINAL GANGLION CELLS (FIRST CERVICAL GANGLION) FROM A CHILD AT BIRTH WITH THOSE FROM A MAN DYING OF OLD AGE AT NINETY-TWO YEARS. (*Hodge.*)

	CHILD AT BIRTH. MALE.	OLD MAN.
Volume of Nucleus	100 per cent.	64·2 per cent.
Nucleoli Visible ...	53 " "	5 " "
Deep Pigmentation	0 " "	67 " "
Slight Pigmentation	0 " "	33 " "

The nuclei have decreased and become irregular in outline. The nucleolus in many cases is not stained by osmic acid, and the cytoplasm is much pigmented, but these cells differ from those fatigued (cat) in that the nucleus does not become darker than the cytoplasm. In the cerebellum also there was an indication that some of the Purkinje's cells had disappeared and others were shrunken.

Still more marked was the effect of age when the nerve cells from the antennary lobe of the honey-bee were examined. In old bees these exhibit a vacuolation of the cytoplasm and shrinkage of the nucleus which makes them quite comparable with the fatigued cells, and very different from those of bees which have just

[1] Hodge, *Journ. of Physiol.*, 1894.

emerged in the perfect form. In this ganglion of the bee the cells actually disappear with age, and Hodge has estimated that for each cell present at senile death there were in the insect just emerged 2·9 cells.

These changes will probably be found more widely distributed in the central system when search is made

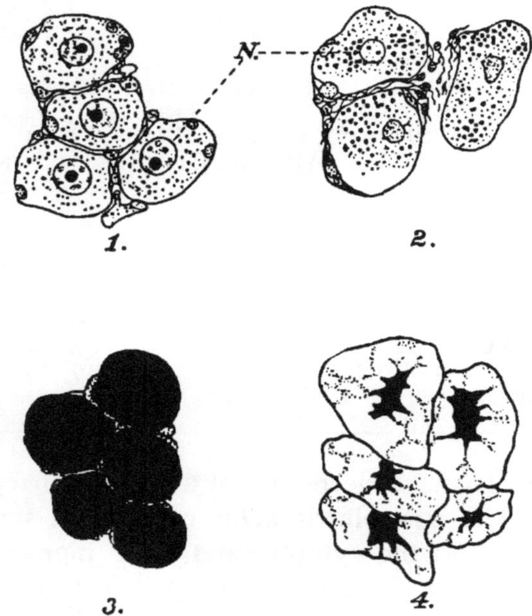

FIG. 77.—To show the changes in nerve cells due to age. 1, spinal ganglion cells of a still-born male child; 2, spinal ganglion cells of a man dying at ninety-two years; *N*, nuclei. In the old man the cells are not large, the cytoplasm is pigmented, the nucleus is small, and the nucleolus much shrunken or absent. Both sections taken from the first cervical ganglion, × 250 diameters; 3, nerve cells from the antennary ganglion of a honey-bee, just emerged in the perfect form; 4, cells from the same locality of an aged honey-bee. In 3 the large nucleus (black) is surrounded by a thin layer of cytoplasm. In 4, the nucleus is stellate, and the cell substance contains large vacuoles with shreds of cytoplasm. (Hodge.)

for them, for pathologists at the present time find the

central nerve cells most responsive to prolonged abnormal conditions of the body even when the disease is located quite outside of this system.[1]

The cause of senile changes in nerve cells cannot be given. Could the central system of an aged person be transplanted to the body of one in the prime of life, possessing a vigorous digestion and perfect circulating system, it might be possible to judge how far such changes are the results of defects in the systems on which the nerve elements depend. Many facts suggest that specialisation in the functions of a cell is the first step towards its ultimate destruction, and in that sense the causes of its death should be mainly inherent. The overgrowth by the supporting tissues and intoxication through defective nutrient conditions are of great importance to this senile dissolution, but their exact values are necessarily unknown. A dependent system is, however, at the mercy of the others with which it is associated, and hence the changes caused by defects in the latter are to be distinguished from those arising in the former. This is especially important when an attempt is made to test the hypothesis that the capabilities of the central system tend to disappear in an order inverse to that in which they have developed. This statement applies to changes resident in the nerve elements themselves. When it is remembered that the last developed cells are smallest, and have but a small quantity of cytoplasm, that the very tardiness of their development indicates their environment to have been less favourable, and finally that those conditions which retard growth also favour senescence, the hypothesis appears plausible. In looking for the anatomical changes in this instance it is to be remembered that there is no proof that the inter-

[1] Schaffer, *Ungarisches Arch. f. Med.*, 1893; Pandi, *Ungarisches Arch. f. Med.*, 1894; Popoff, *Virchow's Archiv.*, 1894.

ruption of a portion of those paths by which the impulses pass from one element to another need cause the complete disappearance of the elements thus disconnected, especially if the break involves only a few of the terminals of the nerve fibre; and so the slow physiological disorganisation of the completed central system is not necessarily accompanied by a corresponding amount of atrophy.

The old age of the central system is in a measure independent of the degree to which it is exercised, unless the exercise be so excessive as to cause continual and extreme exhaustion. So far as known the lumbar enlargement of the sedentary student does not grow old faster than that of a professional runner, and on the other hand there is no evidence to show that the best exercise of the hemispheres does clearly postpone in them the involutionary processes.

It can be easily calculated how many times the heart will beat, the lungs expand, and each of the many bodily functions repeat itself in the course of an ordinary life. Among these functions are the regular changes in the nerve cells. Without doubt this rhythm (recuperation and fatigue) repeats itself each day, and thus all persons undergo alterations of the nervous system as regular as those which occur in stature, but the intensity of this change is an individual matter, and perhaps not fundamentally important as a physiological phenomenon. Viewed from this side, therefore, much truth is contained in the observation that, though the differences between men are slight, yet such as they are, they amount to a great deal.

All through life, one after another, the various capabilities drop away. The power of visualisation is lost, pleasure in music disappears, memory becomes weak, save in narrow lines; a new language, a new science, or a new handicraft appear as very serious undertakings, and as a rule are only indifferently acquired. Yet there can be no doubt that when a capability is lost, then somewhere in the central system a physiological change, if not a structural one, has rendered the cells less active. As behind each functional advance there must be a further structural organisation, so diminished organisation must accompany each loss of power. Of such slow changes in ourselves we are happily for some time unconscious, and the beginnings are to be discovered only by the application of some exact measure of the reaction. In his essay on Old Age, Sir J. Crichton Browne gives some interesting observations bearing on this loss of capability.[1] From these it appears that in England the expertness of the Birmingham button-makers, the Staffordshire potters, and the Bradford weavers increases from the time of entering their trade at about seventeen years up to thirty years of age. From this until forty-five years there is a period of equilibrium, and then a decline sets in. The following table illustrates the manner in which the productiveness of a maker of vegetable ivory buttons, working the same number of hours, decreases with advancing years.

TABLE 62.—SHOWING AT DIFFERENT AGES THE NUMBER OF GROSS PER DIEM PRODUCED BY A SAWYER OF VEGETABLE IVORY BUTTONS. (*Sir J. C. Browne.*)

AGE.	NUMBER OF GROSS PER DIEM.
40 years	100
45 "	80
55 "	60
65 "	40

[1] Sir J. C. Browne, *Brit. Med. Journ.*, London, 1891.

Statistics of this sort are very important, and there should be more of them. At the same time they must be interpreted with caution. The monotony of existence and the general environment of the button-makers is probably conducive to an early decline. Further, this decline may be directly dependent on changes in the spinal cord rather than the cerebrum. Finally the muscles as well as the central system are involved, and there is no collateral proof that other capabilities begin their decline at the same time or continue it at a similar rate. The false inference to be here avoided is the same that was pointed out in connection with the diminution of brain-weight due to age. The conclusions are applicable to other artisans under like conditions, but not necessarily to those living under quite different conditions. Yet there is abundant evidence that the capacity for very high-grade manual work demanding concentrated attention and judgment is limited to a decade or two at the prime of life, and the history of champion athletes points to like limitations.

In this connection the fact that there is a multiple innervation of many parts of the body is perhaps significant. In its simplest form this means, for example, that a given muscle can be contracted in nearly the same way under the influence of different sets of nerve elements. Expertness translated into physiological terms must mean the use of the same elements in the same way to produce a given result. The mechanical precision would be much increased by this, but the strain would always fall on the same elements, which in the case of a manual art are almost certain to be over-exercised. This establishment of a preferred pathway tends to the suppression of activity in the other possible pathways, and to the abolition of that fringe of connections which is normally present in every pathway, but the existence of which, though conducive to the general health of any portion of the central system, is a source of disturbance when great precision is to be attained. Leaving aside the general characters of the last of the "seven ages" of human life which the physicians and poets of all time have united in depicting, there is one aspect of this period, so far as it concerns the nervous system, which may be emphasised. The conditions of this system in old age and in fatigue are closely similar, though the tired cell in the prime of life still possesses the power of recuperation, which in old age is wanting. Even in youth extreme fatigue brings with it the bodily expression of age and the feeling of decrepitude, and in many ways the two conditions thus approximate, although it is recognised that fatigue depends on the exhaustion of stored material and senility on the failure of the power to restore it in full measure, but how the one stage passes over to the other, or why, are problems still untouched.

CHAPTER XVIII.

THE EDUCATION OF THE NERVOUS SYSTEM.

The nerve cells alone educated—Development of central system precocious—Education, natural and formal—The developing system—Anatomically—Psychologically—Aphasia—Basis of intelligence—Limitations of formal education—Discrimination—Fatigue—Memory—Muscular power—Habits—Rhythms—Individual variations—Those of race—Class—Sex—Precocity and the ripening of the faculties—Training first for association—Second for power—The ideal—The double object: vigour and poise.

EDUCATION consists in modifications of the central nervous system. For this experience the cell elements are peculiarly fitted. They are plastic in the sense that their connections are not rigidly fixed, and they remember, or, to use a physiological expression, tend to repeat previous reactions. By virtue of these powers the cells can adjust themselves to new surroundings, and further learn to respond with great precision and celerity to such impulses as are familiar because important.

In its size and development the central system is precocious. Long before birth all the cells destined to compose it are already formed, though by no means all are developed in the sense that they have acquired the form and connections characteristic for those at maturity. At the close of embryonic life the sensory nerves rapidly extend, and the connection of the central cells with limiting surfaces of the body being thus established, all experiences become those of education. The act of living is thus the most important natural educational process with which the human body has to do, yet it is usual to restrict the term education to a series of formal events falling within the period of school life. Formal education as such may have objects which are quite as distinct as those of gymnastics and athletics in the field of muscular training. In their extreme forms these two aims are distinguished by the fact that the athletic or technical training fits the individual to do some one thing which in his day and generation is considered desirable, while the culture or gymnastic method seeks to round him by the better exercise of weaker points, so that his activities may become more complete. In earlier education the culture method, though often disguised, is yet the one mainly pursued, and it is the significance of this method which will here be considered.

In the development of the central system it is found that an anatomical framework is first formed. In this framework are represented, in outline, the nerve structures whose functions are most fundamental. These with later growth are locally strengthened and organised, and by the establishment of associative paths gain both a wider influence and greater complexity of reaction.

In the history of this unfolding of the nervous centres, atavistic tendencies crop out. Most interesting, perhaps, are the prehensile powers of the great toe, and the clinging powers of infants during the first four weeks of life.[1] These capacities, like the sucking reflex, disappear sooner or later, leaving no trace behind, but there can be little doubt that proper examination of the centres concerned would show an histological basis

[1] Robinson, *Nineteenth Century*, 1891.

for the reactions. It is but rarely that a vanishing power can be thus tracked, but there is ample suggestion that many ancestral phases are for a time exhibited and then outgrown by the maturing brain. Among the developing functions sensibility comes first. Indeed the sensory system in certain human monsters may be well developed, although both central and motor elements are quite wanting.[1] To this general sensitiveness, evident early in fœtal life, and without which the central system ceases to be significant, there are rapidly added, about the time of birth or shortly after, the more special sensations of taste, smell, touch, hearing, sight, and of temperature. Then, as organisation progresses, come the emotions of fear, astonishment, anger, closely followed by the development of the intellect and will, with the power of language and self-consciousness. This is not the place to essay the reduction of all these various expressions, intended to indicate grades of mental development dependent on the organisation of the central cells, to their physiological equivalents. Since the special senses first become useful, the various reactions of the individual are customarily associated with one or another of them. That this is somewhat a matter of chance the education of all defectives shows. The development of the parts of the brain devoted to the mediation of the special sensations as well as the sense organs[2] (Luckey) like those of hearing and sight, is rapid at first, then slow, yet continuous up to maturity, as is seen in the curves for cortical growth, already given. With the refinement in sense perception and the accompanying central changes, comes a corresponding increase in the control of the motor elements, and in these latter an increase

[1] Leonowa, *Neurologisches Centralblatt*, No. 20, 1894.
[2] Luckey, *Am. Journ. of Psychol.*, 1895.

in the strength, accuracy, and readiness with which they respond.

Such is the development from the standpoint of the increasing organisation of the system at large. Among the sensory and central constituents are changes of a different sort. In the very young the mental processes are limited by the fact that memory is very poor. As this power increases it becomes possible to hold the mental image for a longer time, until finally a large fraction of the mental operations involves the employment of such images which may be present in the terms of any sense. Thus the child first receives impressions which are quickly lost, then those which are remembered for a time, and finally he accumulates a store of memories which enter as modifying factors into all subsequent mental activity. In advanced age the mental activities have it in common with those of childhood that the images are easily mislaid, but when available during the period of senescence they are more complete than during the earlier stages of growth.

The study of sensory aphasia, designating by this a loss of the power of recognition or expression consequent on injury to the sensory or central elements, has helped much to the understanding of the manner in which our ideas are built up during the formative period. Analysis indicates that the notion of a brass bell is built up from its smell, taste, temperature, weight, shape, colour, and the sound it makes, though the information obtained through the eye and the ear is that on which we commonly rely; but a blind-deaf person would emphasise the other sources of information. When an impression has been received from a bell it may be indicated on our part by a sound: the name or an imitative noise, or a gesture of some sort: a picture or the written word. In the ordinary right-handed person it is the injury to the left cerebral hemisphere between the sensory and motor regions that most often gives rise to symptoms of aphasia. The symptoms vary with the location and extent of the lesion, but depend on the stoppage or diminution of the nerve impulse at some point between its arrival at the cortex and its transfer to the efferent central cells, used for the response. The study of aphasic patients has brought out the fact that any one or more of the several events involved in recognising the bell by means of the principal senses, and expressing this recognition in the usual way, by the written or the spoken word, may drop out. So, for example, the patient may be unable to speak or write the word "bell" when a bell is simply seen, but at the same time be able to do both if it is also rung. Or the sound may elicit only the written but not the spoken word, and so on through the possible combination of sensations and forms of expression.

Thus the impulses reaching an expressive centre, and proving too weak to cause the discharge of it, may still prove efficient if combined with the corresponding impulses from another sensory centre to which the object also appeals. It is therefore the passing of the impulses from all the sensory centres stimulated by the object that gives the basis for the most perfect response. Clinical studies of the kind furnish grounds for the idea that the presentation of an object to any one of the senses revives the mental image of that object in terms of the other senses which may be, and formerly have been, excited by it, and that the more vivid these associated images, the more complete and clear is the conception. As the possibility of forming the extra images is curtailed, the conception becomes weaker, more special, and less reliable.

When the sight of the bell causes it to be named, the

changes in the brain are not duplicates of those occurring when the sound of it has led to the same reaction. In the two cases the same expressive centres have been directly roused by *different* sets of fibres, one from the auditory and one from the visual centres, and the secondary revivals have been correspondingly varied. The bearings of these facts are very wide. Among mammals it is a familiar observation that some, like the rat, are dependent on the sense of hearing, or like the cat, on that of sight, or the dog, on that of smell. This means that the mental images which rule these animals are in terms of the dominant sense, and anatomically that the cortical centres for the dominant sense are those best connected with the motor areas. When men are compared, there are possibilities for very wide differences in these arrangements, the eye rather than the ear being generally the dominant organ. Something of this is doubtless due to training, but probably much more to anatomy. It is interesting to note that in these days of printing a large proportion of our second-hand information reaches us through the eye, while in the earlier centuries the ear was the main channel. Further, owing to anatomical peculiarities an individual may be persuasive, with his pen and yet a doubtful orator. Such every-day combinations may be easily explained. The same sensory portions of the brain are not connected impartially with either the centre for the movement of the hand and arm in writing or that for the muscles of phonation used in speech ; but there may be wide differences in this relation, so that the speaking and the writing man are somewhat different persons. The working value of the mental images appears also as dependent on the number and balance of the secondary sensations which accompany them. The greater the number of these the more certain and precise is our thought. For this reason the development of the intelligence is associated with the perfection of more than one sense organ, whereas reliance on a single avenue of sense, while it may lead to very precise reactions graded in accordance with the intensity of the single sort of stimulus, leaves us without those fringing sensations which form the basis for distinction and comparison.

In biological equations the values of the different factors are often open to wide variation. We have insisted on the three-fold composition of the central system, one factor being afferent or sensory, another central or distributive, and the third motor or efferent; each one of these divisions represented by distinct anatomical cell groups. It is readily seen that a high degree of responsiveness among the central cells gives us the intellectual type of reaction. Where the efferent portion is well organised, we have the anatomical conditions, or the man of affairs or action, while exaggeration of the afferent or sensory component leads to a merely passive existence, or to hysteria, according to circumstances.

Connections between the exercises of formal education and brain change have not been demonstrated. It is not known how a year's schooling affects the central system, and it is not probable that we shall soon arrive at facts of this sort. Available, however, are the facts of anatomical growth during this period, and to these plausible explanations have been given.

The aim at the moment, therefore, is to determine what limitations anatomy places to the educational process, and thus to obtain a rational basis from which to attack many of the pedagogical problems. It appears probable that the education of the schools is but one, and that, too, rather an insignificant one, of many surrounding conditions influencing growth. The average individual is first subjected to some formal training when about three years of age. At this time the number of cell elements is complete, and the history of future organisation has been in its broad outlines determined by their first arrangement. Examination shows that but a fraction of the elements have begun to develop, though growth is everywhere visible, and some of the elements have attained almost their full size. The encephalon at this age has more than two-thirds its adult weight. It is during this last stage of the growing period, which to be sure may continue into the forties, that formal education is called upon to modify the central system. In these pages it has been maintained that this modification might go on without great increase in weight, though with much effect on the organisation, because although in many cases the nerve elements were in the granule stage of development and without prolongations, yet in many others they possessed prolongations which had taken all but the last step necessary to the establishment of full functional connection with neighbouring cells, and hence for such cells the constructional change demanded was very small. The impulse to this change in any part of the system is seen to be the improved circulation, giving a larger nutritive supply, and the direct action of the nerve impulses putting the cells in a chemical condition to best make use of the surrounding nutriment. Such being the conditions, two things must happen. Education must fail to produce any fundamental changes in the nervous organisation, but to some extent it can strengthen formed structures by exercise, and in part waken into activity the unorganised remnant of the dormant cells. No amount of cultivation will give good growth where the nerve cells are few and ill-nourished, but careful culture can do much where there are those with strong inherent impulses towards development. On neurological grounds, therefore, nurture is to be considered of much less importance than nature,[1] and in that sense the capacities that we most admire in persons worthy of remark are certainly inborn rather than made.

Among children there are the widest variations in congenital composition of the central systems, and similarity is neither desirable nor liable to occur save among the members of the same family, or, better still, in cases of twins. It has been made probable that by the cultivating processes of school-training the formed structures tend to be strengthened, dormant elements roused to further growth and organisation, and made more perfect in this or that direction according to the nature of the exercise. By strengthening the formed cells their powers of differential reaction, of organic memory, and resistance to fatigue are increased. By associating given sets of muscular reactions with given sense impressions habits are formed, in consequence of further organisations among the nerve elements, and finally nutritive rhythms associated with the periods of activity and rest are established, with the result of economising the bodily energy, and rendering its expenditure more effective.

In general the power of sense discrimination increases with age, because such discrimination depends mainly on central arrangements which are not elaborated in the earliest years, but where it depends on peripheral arrangements the power may in some cases decrease with age. Thus Czermak [2] has brought forward evidence that the power to discriminate two points on the skin is rather finer in children than in adults, owing to the better

[1] Galton, *Inquiries into Human Faculty*, 1883.
[2] Czermak, *Gesammelte Schriften*, Leipzig, 1879.

innervation of the skin in the former. For as growth proceeds the skin increases in area more rapidly than the nerves which supply it increase in number. In the power of discrimination based on the condition of the peripheral sense organs will probably be found the anatomical condition controlling in large measure the scale of human performance. Here we have in mind the size of things made by man, ranging from the minute and painfully detailed constructions to the broad and carelessly general. Perhaps the extremes in scales in painting and sculpture will best illustrate the point. A Meissonier and Munkacsy could not interchange their styles at will, nor would the style of Michael Angelo suit the needs of an engraver of gems.

While in long periods fatigue in mental operations can be demonstrated, in short periods the results are not so clear, owing in part to the confusion caused by the daily rhythms. Moreover, in the longer tests it is difficult to give proper values to the quantitative reduction in the work done, and at the same time to the decrease in accuracy, since the two do not run a parallel course (Burgerstein,[1] Laser[2]). Ebbinghaus[3] has made the most elaborate tests on memory, and although his experiments were undertaken on himself alone, they suffice to show the easy onset of fatigue and the enormous value of exercise alternating with even lengthy periods of rest. Bolton,[4] who studied in scholars from eight to fifteen years of age the memory span for a series of digits up to nine, found the length of the span as well as the accuracy of its reproduction to increase with age. These improvements in

[1] Burgerstein, *Trans. VII., Internat. Cong. Hygiene and Demog.*, London, 1891.

[2] Laser, *Zeitschrift f. Schulgesundheitspflege*, vii., Jahrg., 1894.

[3] Ebbinghaus, *Ueber das Gedächtnis*, Leipzig, 1885.

[4] Bolton, *Am. Journ. of Psychol.*, 1892.

the central cells find an equivalent modification in the motor elements. Grigoresce[1] examined a series of children of the ages below given, to determine the strength of their grip, as shown by the dynamometer. The figures expressing the power of the grip for each year are in every case averages of a hundred observations.

TABLE 63.—SHOWING THE INCREASE IN THE STRENGTH OF THE GRIP WITH AGE; MEASUREMENTS WITH DYNAMOMETER. (*Grigoresce.*)

AGE IN YEARS.	STRENGTH OF GRIP IN KILOGRAMMES.
7	12·21
8	13·97
9	16·52
10	19·17
11	20·58
12	20·97
13	22·13
14	27·21
15	33·04

To be noted in this table is the remarkable increase during the fourteenth and fifteenth years, equal in amount to that occurring during the previous seven. "The Autobiography of a Strength Seeker"[2] contains a record of most remarkable gains in strength later in life. The case of Windship is certainly exceptional, but is here of interest by reason of the fact that, though these gains are usually a mixed result of changes in the muscle as well as the central nervous system, yet the later in life they occur the more the central system must be held responsible for them, since the muscles soon reach a limit of growth. The capabilities of the nerve

[1] Grigoresce, *Compt. Rend. de la Soc. de Biol.*, Paris, 1891.
[2] Windship, *Atlantic Monthly*, Boston, 1862.

cells are far beyond the limits of their ordinary performance. It will serve to suggest the latent possibilities of this sort resident in the central system if we recall the strength of very moderately developed men suddenly exhibited during attacks of maniacal frenzy.

In all functional activities a tendency to the formation of habits occurs. Concerning these associations there is one feature of much educational importance, generally recognised, to be sure, but which at the same time has been made more clear by the investigations of Bergström.[1] In these experiments the test was to sort as rapidly as possible, and according to suit, a pack of cards, placing them in piles on a table. Another pack, with the cards in a different order, was then taken, and the experiment repeated. This second trial took much longer than the first. The increase in the time was found to be due, not to fatigue, but to certain associations formed during the first sorting, whereby a given suit was connected with a pile having a given location on the table. In the second pack the order of the cards was of course different; the suits on the table were therefore differently located, and the memories of the first associations still persisting, they directly interfered with the second performance, causing false movements, and so increasing the time. The demonstration here of the loss of energy in learning what needs only to be unlearned is very striking, and if one experience produces such an effect it is not difficult to understand how habits early formed and long cultivated become so difficult of eradication. With habits come rhythms in activity, and, recognising the importance of rhythms of this nature, the training may be adjusted to them, thus catching the system at the most favourable moments.

[1] Bergström, *Am. Journ. of Psychol.*, 1893.

In this history of organisation there are many errors, happily for the most part transient. Here belong all those vague suggestions of a second personality which in growing children are so perplexing and so numerous. Sometimes these phases periodically recur,[1] and at adolescence they are most accentuated. They represent the disharmony of the first steps in change and progress, and, like mirror-script, gibberish talk and the various forms of general inanity are but intermediate phases between the imbecility of the infant and the intelligence of the adult.[2] The intensity with which any form of exercise is carried on during the growing period leaves its trace, and the absence of it at the proper time is for the most part irremediable. We should hardly expect much appreciation of colour in a person brought up in the dark, however good his natural endowments in this direction. Thus any lack of early experience may leave a spot permanently undeveloped in the central system—a condition of much significance, for each locality in the cerebrum is not only a place at which reactions, using the word in a narrow sense, may occur, but by way of it pass fibres having more distant connections, and its lack of development probably reduces the associative value of these also.

It is now recognised that thought can be carried on in terms of the several senses. In this connection Fraser[3] has made an examination of certain philosophic writers which indicates that particular writers or schools prefer sense-images of one mode in their speculative thought, and he suggests that much of the failure to be mutually comprehensible, depends on the fact that

[1] Siegert, *Die Periodicität in der Entwickelung des Kindesnatur*, 1891.

[2] Clouston, *Neuroses of Development*, 1892; Warner, *Physical Expression* (International Scientific Series), 1893.

[3] Fraser, *Am. Journ. of Psychol.*, 1892.

tactual and visual images, for example, are by no means capable of being manipulated in the same manner, and hence that relations conceivable in the terms of one are often not so in those of the other. With the employment of one sort of mental image comes precision, but it is precision gained at the price of limitations. Fortunately the law of the diffusion of incoming impulses works against a too great specialisation in this direction. Yet in the highly defective this specialisation must be carried very far, and in those whose endowments are distinctly unusual the dominance of one sense in controlling the reactions of the central system may rise to the dignity of a deformity. It would be a pallid truism to insist that persons are very different in the anatomy of their nervous system, and also in their nervous activities, were it not that educational practice appears so often to start from quite contrary assumptions.

The description of racial differences in nervous reactions forms a literature by itself, and the traits of widely separated classes in the same community are almost as different as those of unlike races. In Bavaria Ranke [1] found the cranial capacity of the townspeople distinctly greater than that of the peasants in the surrounding country—a relation which probably means that in general the better endowed individuals sought the severer struggle of the town with its greater excitements and compensations. It is to be anticipated that one great difference in races will be found to lie in the extent of growth and organisation in the nervous system after birth, and especially after puberty. Should it turn out on further examination that some of the lower races lose their capacity for later training after adolescence, we should look with interest for the changes in

[1] Ranke, *Der Mensch.*, 1894.

the cerebral cortex in order to determine whether growth there practically ceased at puberty; for, by contrast, Venn, studying the size of the head in Cambridge students, found it on the average greatest and growing for the longest time in the group of most successful men. The accomplishments of this fortunate group are therefore to be associated with innate capacities, and have small ethical significance; they may be admirable, just as are the paces of a well-bred colt, but the colt deserves no credit for its gait. Indeed the observations of Ranke just quoted, stand as an anthropological verification of that tendency so to use our native endowments as to get from them all the stimulation of which our sensory systems are capable. It is a deeply seated impulse. The defective sense organ or the paralysed limb are often treated very roughly in the vain hope of getting a sensation from them. It would appear that the parts of the nervous system cut off from their normal stimuli may, so long as they are alive, cause their owners much discomfort, and these individuals resort to indirect means to get rid of the irritation.[1] The obverse of this appears in the "breaking out" experiences so common in all communities of growing individuals. In prisons, schools, and similar institutions, where hard physical exertion is not compulsory, or the routine is deadly monotonous, such explosions are bound to occur. Some years ago, when college athletics were coming into vogue, Professor Richards, of Yale, pointed out that college rows and disorders fell off as athletics came in.[2] It is not easy to make experiments of this sort, but there is no doubt that during the growing period this surcharging, though

[1] Weir-Mitchell, *Injuries of Nerves and their Consequences*, Philadelphia, 1872.

[2] Richards, *Pop. Sci. Month.*, 1884.

not regular, is an event to be expected, and may take almost any form of expression.

In this connection the consideration of the problems of education, as modified by sex, forms an important topic. While from the anthropological standpoint there is a typical man and typical woman for each race, these are not the same for different races. In the secondary sexual characters there are some distinctions of general applicability—for instance, women are on the average smaller than men. Stature and weight are, together with proportion, the best marked secondary characters by which the sexes are distinguished, and yet these overlap in every way. Among such secondary characters is that of the nervous system, and there we find a similar overlapping. There is no question about the fact that women have on the average smaller brains, though the record from a better class of women than those furnishing the data now employed would perhaps raise the average, but these in turn must be compared with records from a better class of men. This small absolute weight is in no wise mitigated by the fact that the weight of the brain as compared with the weight of the body is greater in women than in men, for, as we have seen in earlier chapters, if that were a criterion, we should all bend before the massive intelligence of the new-born child, whose proportional brain-weight is six times greater than that of the adult. The suggestion has been made that the female brain is lighter because its structural elements are smaller. Granting this, the significance of the absolute size of the elements still requires to be explained. The only interpretation that we have for the size of these elements is as an expression of the power to store and discharge force in a short period of time, and to furnish branches for structural connections. Such a brain of small elements, no matter to which sex it belongs, has the same characters, but so complicated are the reactions of it with nutritive conditions that any inference from mere size has little value. If the influence from size were applied thoroughly, mental superiority would reside with the tall as contrasted with the short men, since as a rule tall men have the heavier brains. Size, therefore, has a meaning, but is by no means entitled to dominate the whole interpretation of the central system. There is little or nothing in the weight relations of the female encephalon to show it different from that of the male. In reactions, however, the female has a more local responsiveness than the male, and back of all this is the matter of general physiology, which has its distinct modifications according to sex. Moreover, it is impossible to escape the conclusion that in women natural education is completed only with maternity, which we know to effect some slight changes in the sympathetic system and possibly the spinal cord, and which may be fairly laid under suspicion of causing more structural modifications than are at present recognised. Basing the inference on the size of the structural elements, we should infer that the typical central system in the female would be somewhat more easily fatigued, and also be slightly less complete in organisation.

For the rest we have no anatomy, and only modes of response, on which to base a judgment. The characteristic reactions seem in part to depend on fundamental physiological differences, in part on the organisation of the central system, and in part on nurture. Just how far any variations in this last condition will modify the reactions of women is at present a matter of experiment, an experiment in which the brain-weight question cuts a much smaller figure than was at one time imagined.

In all cases the process of education must be much

influenced by mental ripening. The child may be precocious or backward. It is interesting to note, apropos of what has been said concerning strength and size, that precocity is for the most part concerned with an early increase in the complexity rather than the strength of the reactions; it is a precocity of organisation, not of size. Moreover it is apt to appear along limited lines only. Physiologically it is growth without the usual sensory provocation, and as such renders superfluous much of the formal training, the purpose of which is to stimulate. Hence the ease with which these individuals may learn some things. The best studies on the subject show that precocity and genius go together.[1] The same conditions which gave the individual a generously planned nervous system also favour its early development. In such precocious persons it continues to grow for a longer period than usual—a feature which is fully as important as the precocity itself.[2] It is extremely interesting to see how in a series of eminent men, excluding men of action, the determination of distinction follows the order in which the brain normally attains the high development necessary to command recognition in a particular profession.

The manner in which the encephalon becomes organised shows that some combinations among the central elements are normally completed earlier than others. When, however, such sensori-motor connections are once established, the time required for their perfection is comparatively short. Further, it appears that high excellence, especially in the acts requiring simple sensori-motor combinations, is acquired speedily or not at all; and thus prolonged exercise, though it may have value as a moral training, is insignificant for technical improvement.

[1] Sully, *Pop. Sci. Month.*, 1886. [2] Galton, *Hereditary Genisu*, 1884.

TABLE 64.—SHOWING, FROM 287 CASES OF MEN DISTINGUISHED IN THE PROFESSIONS NAMED, THE PERCENTAGES OF THOSE WHO GAVE PROMISE BEFORE 20 YEARS, PRODUCED BEFORE 30 YEARS, AND WERE DISTINGUISHED BEFORE 40 YEARS.

The order is the same in all three periods, except in the case of the Scholars, who, though giving early promise, were late in production and in attaining distinction. (Sully.)

PROFESSIONS.	GAVE PROMISE BEFORE 20 YEARS OF AGE.	PRODUCED BEFORE 30 YEARS OF AGE.	DISTINCTION BEFORE 40 YEARS OF AGE.
Musicians ...	95	100	100
Artists ...	89	98	100
Scholars ...	83	**71**	**90**
Poets ...	75	92	92
Scientists ...	75	80	92
Novelists ...	75	56	80
Philosophers	67	56	60

Those professions demanding only small acquisition, but a very perfect adjustment between one sense organ and one set of muscles, as between the hand and the ear in the musicians, and the eye and hand in the artists, are precocious throughout, while the philosophers with their need for accumulated information and ripened judgment bring up the rear. Similar investigations on slightly different material yield accordant results.[1]

The precocity which is so marked in the formation of the structural elements, and the slowness with which they complete their organisation and development, are features fundamental for the production of a good intelligence, for the first condition supplies the elements for response to the various stimuli from without, and the second prevents the formation of reactions too readily organised before the frequent repetition of a

[1] Elliott, *Internat. Rev.*, 1882.

stimulus has given assurance of its importance. From such facts the general relations of formal education to the growing process are fairly evident. The function of it is to round out the original framework of the central system, in accordance with the natural provisions there present. Without question there is something very fatalistic in this. No amount of education will cause enlargement or organisation where the rough materials, the cells, are wanting; and on the other hand, where these materials are present, they will, in some degree, become evident, whether purposely educated or not. Anatomically, the process of training leads to organisation, extending from the centre first developed, so that regions isolated in infancy become later more closely associated. So, too, it is found that on the motor side, the control of a limb by the brain is first established over the limb as a whole, since the cells controlling the joints nearest the trunk mature earliest.

Since the cortical centres for the more proximal joints tend to be most speedily organised, they become, by virtue of this organisation, in some degree a thoroughfare for the impulses to the other cortical centres controlling the more distal groups of muscles. Thus the impulses tend to pass both through the cortex and down the limb in regular series.[1] Seguin thus found it most advisable to begin the training of an idiot hand with movements at the shoulder, and the later brain physiology as well as the history of growth confirms his method. The suggestion from this is that the exercise of those structures best formed will most readily arouse to activity those next to be developed, and that direct exercise should follow close upon the natural extension of the growth processes.

In any special case it is hardly possible to predict

[1] Seguin, *Archives of Medicine*, 1879.

what capabilities may be latent, and earlier education thus resolves itself into a reconnoissance among the nerve centres for the purpose of finding those which will best act together. By consequence the training for association comes first, just as precocity tends to first take that form, that for dexterity and strength being deferred until later. But all growth is accompanied by impulses to activity, due to the surcharging of the central system, and these express themselves sometimes as disorderly outbreaks, sometimes as enlarged susceptibilities.

Since the beginning of the educational experiment at the dawn of civilisation the problem has been to rouse an interest in these formal exercises of the schools. From the physiological side, that which rouses an interest tends to quicken the pulse and determine a full blood supply to the entire central system, yet the narrow gymnastics of the school, in the most austere form, do not in themselves produce that condition of good nutrition favouring the best diffusion of the impulses and the formation of secondary and subconscious associations. The problem, therefore, of an emotional background has been met with a hickory stick and with gold, as well as everything between them that could be considered as a general excitant; but useful as all these methods sometimes are, it has in the main only been successfully met by a more subtle sympathy and knowledge—a sympathy which in one way or another discovered the growing point in the child and fitted the task to the necessities of the individual.

There remains another aspect of the subject before closing this partial review. The avowed aim of certain educational schemes is to produce a rounded, balanced individual as an outcome of the training process, a psychological result comparable with the ideal human form at one time sought in sculpture. Since conditions of life on the globe are not uniform, and since man only approaches the ideal in his development when in harmony with his surroundings, such a universal ideal is as fanciful as was the notion of Goethe concerning the "Urpflanze"; a sort of grandfather of all the plants possessing the characters of its multiform descendants, yet displaying them with an ancestral simplicity worthy of the golden age of which it had formed a part.

As a matter of fact, the education of an individual is a very local problem in its details. The weak points in the central system must be strengthened, that the abilities given by the strong ones may be guided by some sort of balanced judgment. But the balanced and judicial states are, so far as they go, plainly statical, and the vigour of a healthy restlessness is very necessary if there is to be advance. While growth continues, things bodily and mental are lop-sided, for growth is never general, but accentuated, now at one spot, now at another. But this very unbalance, if only it be the outcome of natural endowment and not of *a priori* training, gives a vigour not otherwise to be obtained.

The history of the normal individual is through various phases of unstable equilibrium and awkward strength, to the poise and quiescence of late maturity, yet in any community examples of all these phases are found as terminal states in both old and young. The formal methods, therefore, which shall recognise, in the presence of these enormous differences in endowment, the dynamic value of the natural inequalities of growth, and utilise them, preferring irregularity to the roundness gained by pruning, will most closely follow that which takes place within the body, and thus prove most effective.

CHAPTER XIX.

THE WIDER VIEW.

The brain the organ of the mind—Civilisation and brain-weight—Stability of the central nervous system—Variations in mental power during historic times—The education of ancestors—Civilisation and the subdivision of labour—Efficiency of modern effort—Native mental power—Legitimate aims in education—Direction of training—The background of growth.

IT is a familiar idea that mental performance is accompanied by nervous change. Could the relation between these two sets of events be adequately explained, at least one fundamental question would be answered. Yet to this fatuous task students of science are no longer attracted, and the method of the day is to attack such problems in detail and indirectly. Thus have grown up provisional hypotheses, and present controversy centres about the degree of support which can be given to these competing explanations.

Let the explanation, however, be as it may, the prime fact remains that the phenomena of consciousness are exalted or depressed by purely physical conditions; that they unfold, flourish, and fade within a single lifetime, and may be blighted or become monstrous by misuse or by disease. This dependence of the mind upon the brain is certainly most striking, and it has naturally suggested that, by the study of the body considered as a complex machine, the beginnings and relations of mental phenomena could in some cases be discovered. Yet by this effort we touch our problem on one side only, and so must be prepared to meet modifying facts which are certain to come in from other sources.

The problem of education in its most general form demands an answer to the question, how the individual is to make the best use of his own limited life-cycle, while at the same time those larger responsibilities of the unit to the race are kept in mind.

The facts which may be brought forward as materials for the reply are, however, very baffling.

A study of the foregoing tables indicates that races which have progressed farthest in civilisation are also those which possess a large brain-weight; but the converse of this proposition is by no means true, for the tables also show that there are races possessed of a large brain-weight and yet uncivilised. It would be hardly fitting to introduce here matters which have taxed the best ingenuity of the anthropologists, but it is worth while, perhaps, to make some effort towards the clear expression of the fact that so-called civilisation is not necessarily accompanied by an increase in the native mental power of the individuals who take part in it.

These difficulties have recently been very concisely presented by Boas,[1] who concludes that races low in the scale may still have before them a future of development, and suggests certain methods for the study and determination of race characters.

The turn of any race towards progress depends, like all other reactions between living things and their environment, upon a series of happy accidents. It must

[1] Boas, *Human Faculty as determined by Race.* Address before the Section of Anthropology, Am. Ass. Ad. Sci., August, 1894.

be remembered that, as a rule, our latent capacities as individuals are far beyond our regular achievements, and that the stimuli which shall bring these powers into action may be of very different sorts. An isolated race may apparently have progressed but little, when suddenly an earthquake, an invasion by a neighbour, or the birth of a more generously endowed member of the community, serves to give them an impulse which may not be exhausted for many generations.

Striking indeed are the different ways in which such communities respond to the new influences brought to bear upon them from without, grappling at one and the same moment with novel activities, ideals, and diseases. Wallace, in all his journeyings through the Malay Archipelago, was impressed by the differences in the degree with which communities, apparently similar, responded to European influences. Thus far, however, it has been impossible to state the anthropological peculiarities of those who have supported such a trial, as contrasted with the characters of those who have succumbed to it. Further, the question is still to be answered, whether the exercise of the nervous system demanded in highly civilised societies really causes an enlargement, as the exercise of glands and muscles can be shown to do.

The comparison here suggested must be made with caution ; glands and muscles are of slow growth, in one sense of slower growth than the central system. They shrink when the body is starved, and expand when it is well nourished. The central nervous system seems more stable, and starvation at least appears to produce little or no change in its bulk. Such powers of resistance probably have their shadow side, and the system which is thus but little affected in its bulk by unfavourable conditions may be equally resistant to improvements in its environment, as represented by better exercise and more generous nourishment.

It is a fair question to ask, though one hardly possible to answer, whether the best capacities of the best men have increased within historic times. We should hope that they had, and yet a demonstration is at present impossible. But if this is so, and if these exercises leave no structural modifications to the advantage of the next generation, what then is gained by painful study and mental gymnastics throughout a laborious lifetime?

The statement in this form, however, begs the question, for the enormous difficulty of measuring differences of human faculty makes it quite impossible for us to record slight changes, thus leaving the question really in the balance. In forming an opinion, the fact should not be lost from sight that education by rational methods and applied on a large scale is a very modern achievement. Results, too, may be attained quite aside from increase in the number of the nerve cells or marked alterations in their size.

The individual stands related to the world of ideas about him as a race previously isolated does to the civilisation brought to it from without. In some cases such a race may in a few years assume the culture of its visitors, yet it can hardly be supposed that this is accompanied by corresponding anatomical changes on the part of the imitators. It is rather that they find in their visitors a new stimulus, a variation in their surrounding conditions able to excite latent powers which are at that time prepared to be aroused. In the same way the education of one generation forms a more favourable environment for those in the next, and so the efforts of those who precede return to the advantage of those who follow. The possibility, however, of this utilisation of previous exertion exists only where the machinery of civilisation is developed, for the advantages which the descendants enjoy are mainly expressed by relief from unnecessary and wasteful toil and life in a more invigorating atmosphere.

The effective power of a community, depending as it does upon the subdivision of the labour within it, and the mutual dependence of one part on the others, and its trust in them, alone renders it possible for any group of individuals to attend to its own affairs with the necessary freedom from distraction. Individualism, in the biological sense of self-sufficiency, is the one necessary condition for *uncivilised* existence, and so long as each member of the community is concerned with the preparation for all emergencies, just so long is he debarred from advance.

If in any science a student were called upon to develop his subject in theory and technique from the very beginning there is no doubt that such a one would acquire much useful mental exercise, but the body of fact that to-day forms each branch of knowledge, and which furnishes ever-increasing data upon which the logic of succeeding generations is to be exercised, this body of fact, the stored efforts of preceding workers, would be largely lacking, and we should be limited to just so much of science as could be obtained by one man in a long lifetime.

The degree of interdependence is in some sense a measure of the degree of civilisation, but because it has proved marvellously useful to store in an available form the results of the activity of our predecessors, it by no means follows that by this act our own mental powers have attained a greater vigour or a broader span. As well measure the diligence of two farmers by the size of their crops, when one has planted good seed in fertile land, and the other poor seed in arid soil, as measure the mental activities of the members of two communities by the actual contributions of these communities to knowledge, without taking into account the background of their respective cultures.

To be sure, the logical processes of the civilised man seem to count for more to-day than in the past, yet the first forms of tools, weapons, and the arts were no paltry achievements, and a discovery as epoch-making as the potter's wheel, the saw, or the bow would be an enviable distinction for any man, however recent was his birth. Yet by our common measure these great inventors were all uncivilised.

In comparing, therefore, remote times with the present, or in our own age races which have reached distinction with those which have remained obscure, it is by no means clear that the grade of civilisation attained is associated with a corresponding enlargement in the nervous system, or with an increase in the mental capabilities of the best representatives of these communities. Certainly the members of the community at large find themselves surrounded by conditions which favour mental growth just as they are surrounded by conditions which protect them from accident and from disease; but length of human life has not thereby been increased. Eighty or more years is long to live, as it always has been since historic time began, and that more men live out their allotted days in these later centuries, is due rather to a change in the conditions which surround them than to any alteration in their natural powers. Thus our generation makes great marches into the unknown; enabled to do this, not so much because the living are a superior race, but because civilisation has, as it were, constructed military roads throughout the country, and thus the "going" is easier. The physiological hindrance to increase of function seems to lie in the difficulty of effecting a recognition by the nervous system of the stimuli that act upon it. The world about us is not so different for those in the same community, yet how unlike are the responses! Further, when an individual more perfectly constituted has been penetrated by some new sensation, how efficient such a one becomes in awakening in his neighbours similar responses, and thus putting into the possession of the community at large that which was previously the experience of one alone!

Herein lies the significance of those members of society who deviate from the average of their fellows: these deviations are in all directions,—conservative and destructive,—and while not both have an equal chance of perpetuity, they both have great power to modify communal life.

To-day, as in the past, it is permitted to a member of the great majority to grasp only a fragment of the knowledge of his time, with a suggestion of its past connections and something of its present bearing, and to illumine this with so much of foresight as a parsimonious fate has meted out to him. To improve these powers the effort must be made in that direction where response is most ready, and so the formation of habit and reduction of mental friction, by means of concentration, must ever remain the chief objects of a formal training.

Wider knowledge makes us accessible to a larger number of ideas, but it is a preparation for occasional stimuli, rather than for the continuous difficulties, while the manner of reaction, the control and method of performance, are the means by which the welfare of a community at large is most directly influenced through educational endeavour.

The records of enthusiasts indicate that from time to time the hope has been entertained that through formal education much benefit would come to all, and fundamental changes be wrought in man himself. No one questions the good which has followed from this effort, but how far the human organism has been modified by the experience is the doubtful point. Knowledge comes, for the hindrances to knowledge are in a large measure from without, but wisdom, as heretofore, continues to linger, and still to occupy its place as the rare performance of a balanced brain. We feel, though it must be granted that it is a feeling merely, that the descendant of several generations of educated ancestors should have a nervous system favourably modified, more vigorous, more responsive, more accurate in its reactions, and growing, perhaps, for a longer time, thus extending the period of its adaptability. But for this the evidence must still be sought.

As an aid in the solution of these problems, I have endeavoured to combine the various observations on the changes in the form and functions of the brain during its time of growth, believing that a closer study of these changes will undoubtedly assist us, and that along this line the anatomist and the psychologist can bring together their accumulated facts for the benefit of all.

INDEX OF AUTHORS.

INDEX OF SUBJECTS.

www.ingramcontent.com/pod-product-compliance
Lightning Source LLC
Chambersburg PA
CBHW021940220326
41599CB00011BA/942
9783744723800